ELECTRICIAN'S
Technical Reference

Hazardous Locations

Online Services

Delmar Online
To access a wide variety of Delmar products and services on the World Wide Web,
point your browser to:
        http://www.delmar.com/delmar.html
        or email: info@delmar.com

thomson.com
To access International Thomson Publishing's
home site for information on more than 34 publishers
and 20,000 products, point your browser to:
        http://www.thomson.com
        or email: findit@kiosk.thomson.com

A service of **I(T)P**®

ELECTRICIAN'S
Technical Reference

Hazardous Locations

Richard Loyd

Delmar Publishers

an International Thomson Publishing company I(T)P®

Albany • Bonn • Boston • Cincinnati • Detroit • London • Madrid
Melbourne • Mexico City • New York • Pacific Grove • Paris • San Francisco
Singapore • Tokyo • Toronto • Washington

Notice to the Reader

Publisher does not warrant or guarantee any of the products described herein or perform any independent analysis in connection with any of the product information contained herein. Publisher does not assume, and expressly disclaims, any obligation to obtain and include information other than that provided to it by the manufacturer.

The reader is expressly warned to consider and adopt all safety precautions that might be indicated by the activities herein and to avoid all potential hazards. By following the instructions contained herein, the reader willingly assumes all risks in connection with such instructions.

The publisher makes no representation or warranties of any kind, including but not limited to, the warranties of fitness for particular purpose or merchantability, nor are any such representations implied with respect to the material set forth herein, and the publisher takes no responsibility with respect to such material. The publisher shall not be liable for any special, consequential, or exemplary damages resulting, in whole or part, from the readers' use of, or reliance upon, this material.

Delmar Staff
Publisher: Alar Elken
Acquisitions Editor: Mark Huth
Developmental Editor: Jeanne Mesick
Production Manager: Larry Main
Art Director: Nicole Reamer
Editorial Assistant: Dawn Daugherty
Cover Design: Nicole Reamer

COPYRIGHT © 1999
By Delmar Publishers
a division of International Thomson
Publishing Inc.

The ITP logo is a trademark under license.

Printed in the United States of America

For more information, contact:

Delmar Publishers
3 Columbia Circle, Box 15015
Albany, New York 12212-5015

International Thomson Publishing Europe
Berkshire House 168-173
High Holborn
London, WC1V 7AA
England

Thomas Nelson Australia
102 Dodds Street
South Melbourne, 3205
Victoria, Australia

Nelson Canada
1120 Birchmount Road
Scarborough, Ontario
Canada, M1K 5G4

International Thomson Editores
Campos Eliseos 385, Piso 7
Col Polanco
11560 Mexico D F Mexico

International Thomson Publishing GmbH
Konigswinterer Strasse 418
53227 Bonn
Germany

International Thomson Publishing Asia
60 Albert St.
#15-01 Albert Complex
Singapore 189969

International Thomson Publishing—Japan
Hirakawacho Kyowa Building, 3F
2-2-1 Hirakawacho
Chiyoda-ku, Tokyo 102
Japan

1 2 3 4 5 6 7 8 9 10 XXX 03 02 01 00 99 98

Library of Congress Cataloging-in-Publication Data

Loyd, Richard E.
    Electrician's technical reference : hazardous locations / Richard
Loyd
        p.   cm.
    Includes index.
    ISBN 0-8273-8380-0
    1. Electric engineering—Safety measures.   2. Explosionproof
electric apparatus and appliances.   3. Commercial buildings–
–Electric equipment—Safety measures.   I. Title.
TK152.L67   1999
621.319′24′0289—DC21
                                                      98-44676
                                                      CIP

Dedication

I wish to dedicate this book to my wife, Nancy, who has given me so much encouragement and assistance throughout my career. She's my associate, my girl Friday, my number one, and my biggest fan. Without her, this book as well as the other books, would never have been possible. Thanks, Nancy, I love you.

I would also like to dedicate this book to some of the outstanding electricians that I have learned so much from during my career as an apprentice and student and later as a young journeyman electrician. I have continued to learn from others throughout my career as a contractor, inspector, and code administrator. Today I continue to learn from my peers and students; thank you all. Special thanks go to Earl Featherston, Rub Cragan, Blaine Hogue, Red Allen, Dale Nieman, and Dean Evans, all in Pocatello, Idaho, IBEW Local Union 449; D. Harold Ware, of Oklahoma City; Bill Harris, Cecil Harrell, Clifford Vann, Bill Taylor, George Ingram, and Larry Sharren, all of Little Rock, Arkansas; and Ray Olson, Artie Barker, Gary Gould, and Tom Swinehart (Tom taught me the importance of the arrangement and use of words) of Boise, Idaho. I was reluctant to list these names because there were so many more that helped me get to where I am today. I never met a person who couldn't teach me something, and I have gained some of my most prized knowledge from some of the most unlikely sources. I can honestly say thanks to everyone I ever met.

Contents

Foreword

Electric Per Se   Electricity! In ancient times it was believed to be an act of the gods. Neither Greek nor Roman civilizations understood this thing: electricity. It was not until about A.D. 1600 that any scientific theory was recorded when after seventeen years of research William Gilbert wrote a book on the subject titled *De Magnete*. It was almost another 150 years before major gains were made in understanding electricity and that it might some day be controlled by man. At this time Benjamin Franklin, the man we often refer to as the grandfather of electricity or electrical science, and his close friend, the great historian Joseph Priestley, began to gather the works of the many scientists who had been working independently world-wide over the many years. (One has to wonder where this wondrous industry would be today if the great minds of yesteryear had had the benefit of the great communication networks of today.) Benjamin Franklin traveled to Europe to gather this information, and with that trip our industry began to surge forward for the first time. Franklin's famous kite experiment took place about 1752. Coupling the results of that with other scientific data gathered, he decided he could sell installations of lightning protection to every building owner in the city of Philadelphia. He was very successful in doing just that! Franklin is credited with coining the terms *conductor* and *nonconductor*, which replaced *electric per se* and *non electric*.

Figure F–1   Electricity is here! (Courtesy of Underwriters Laboratories.)

Figure F–2 1893 Great Colombian Exposition at the Chicago World's Fair. The Palace of Electricity astonished crowds—it astonished electricians too, by repeatedly setting fire to itself. (Courtesy of Underwriters Laboratories.)

With the accumulation of the knowledge gained from the many scientists and inventors of the past, many other great minds emerged and lent their names to even more discoveries. Their names are familiar as other terms related to electricity. Alessandro Volta, James Watt, Andre Marie Ampere, George Ohm, and Heinrich Hertz are names of just a few of the great minds that enabled our industry to come together.

The next major breakthrough came about only 100 years ago. Thomas Edison, the holder of hundreds of patents promoting DC voltage, and Nikola Tesla, inventor of the three-phase motor and the promoter of AC voltage, began an inimical competition in the late 1800s. Edison was methodically making inroads with DC current all along the eastern seaboard, and Tesla had exhausted his finances for his experimental project with AC current out in Colorado. The city of Chicago requested bids to light the great "Colombian Exposition of 1892." To light the Chicago world's fair, which celebrated the 300 years since Columbus's discovery of America, would require building generators and nearly a quarter of a million lights. Edison teamed up with General Electric and submitted a bid using DC, but George Westinghouse and Nikola Tesla submitted the lowest bid using AC and were awarded the contract. It is said that thousands came out to see this great lighting display. But almost as spectacular as the lighting were the arcing, sparking, and fires started by the display. Accidents were occurring daily! It is said that Edison was embellishing the dangers of AC current by electrocuting dogs while promoting DC current as being much safer. But the public knew better since plenty of fires and accidents were also occurring with DC current.

 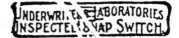

Figure F–3   Examples of early equipment labels. (Courtesy of Underwriters Laboratories.)

About this time, three very important events took place almost simultaneously. The Chicago Board of Fire Underwriters hired William Henry Merrill, an electrician from Boston, as an electrical inspector for the exposition. Mr. Merrill saw the need for safety inspections and started "Underwriters Laboratories" in 1894. In 1881 the National Association of Fire Engineers met in Richmond, Virginia, and drafted a proposal that became the basis for the *National Electrical Code®*. In 1895 the first nationally recommended code was published by the National Board of Fire Underwriters (now the American Insurance Association). In 1897, using this 1895 code as a basis, the first *National Electrical Code®* was drafted through the combined efforts of architectural, electrical, insurance, and other allied interests. The first *National Electrical Code®* was presented to the National Conference on Electrical Rules composed of delegates from various national associations who voted to unanimously recommend it to their respective associations for adoption or approval. And a group from the city of Buffalo, New York, hired Westinghouse and Tesla to build AC generators for Niagara Falls; AC current had won out over DC current. The rest is history!

Preface

This book is written in recognition of the fact that the electrical field has many facets and that the user might have diverse interests and varying levels of experience. Therefore, the interests of the electrical designer, the consulting engineer, the installing electrician, and the electrical inspector were carefully considered in this book. Here is a ready source of basic information on the design of the most common hazardous locations for industrial, institutional, commercial, and residential occupancies. This book can be used by engineers, contractors, and electricians as a reference and can serve as an aid in selecting the appropriate wiring methods for hazardous location installations. It will provide students specializing in industrial and commercial electrical power and control systems with a concise, easily understandable study guide of the classification and installation guidelines for hazardous (classified) locations, electrical wiring methods, and basic electrical design considerations for hazardous (classified) locations. This book is also useful for examination preparation (its primary purpose) and as study material for students and electricians.

This book is brief and concise, allowing easy application to everyday practical problems, and is supplemented with pictures, drawings, examples, and tables, many of which are extracted from the 1999 *National Electrical Code* (*NEC®*). The information facilitates efficient, safe, and economical application of the various types of available wiring methods.

For more detailed information on the specific needs of users; reference should be made to the current edition of the *NEC®,* other National Fire Protection Association related standards, engineering textbooks, application handbooks, and appropriate manufacturers' catalogs.

Acknowledgments

Allied Tube & Conduit Company
Appleton Electric Company
American Iron & Steel Institute (AISI)
Picoma Industries Inc.
Bussmann Manufacturing Company
Crouse-Hinds Division of Cooper Industries
Carlon, a Lampson & Sessions Company
OCAL INC.
Perma-Cote Industries
Pyrotenax USA Inc.
Red Dot L.F. Mason Co.
Robroy Industries Inc.
Square D Company
R. Stahl Inc.
Triangle Wire & Cable Company
B-Line Systems Inc.
LTV Steel Tubular Products Company
Wheatland Tube Company
Western Tube
Underwriters Laboratories

Delmar Publishers and the author wish to acknowledge and thank the members of our editorial panel for their help in developing the concept of this series and reviewing the manuscript.

A.J. Pearson, Director
National Joint Apprenticeship and Training Committee
Upper Marlboro, MD 20772

Robert Baird
Independent Electrical Contractors Association
Alexandria, VA 22314

Brooke Stauffer
Bethesda, MD

P.C. Paul Howard
LaPorte, TX

Kevin Early
Erie, PA

Jeffrey Lew
Purdue University
West Lafayette, IN

About the Author

Richard E. Loyd

Richard E. Loyd is a nationally known author and consultant specializing in the *National Electrical Code®* and (*NEC®*) model building codes. He is president of his own firm, R&N Associates, located in Perryville, Arkansas. He and his wife, Nancy, travel throughout the United States presenting seminars and speaking at industry-related conventions. He also serves as an *NEC®* expert at 35 to 40 meetings per year of the International Association of Electrical Inspectors (IAEI) throughout the United States. Mr. Loyd represents the Steel Tube Institute of North America (STI) as an *NEC®* consultant. He represents the American Iron and Steel Institute (AISI) on ANSI/NFPA 70 of the *NEC®* as a member of Code Committee Eight, the panel responsible for raceways. He is currently the immediate past chairman of the National Board of Electrical Examiners (NBEE).

Mr. Loyd is actively involved in conducting forensic inspections and investigations on a consulting basis and serves as a special expert in matters related to codes and safety. He is vice chairman of the Electrical Section of the National Fire Protection Association (NFPA) and a member of the Arkansas Chapter of Electrical Inspectors Association (IAEI) executive board. He is a contributing editor for Intertec Publications (EC&M Magazine).

Mr. Loyd is an active member of the IAEI; the NFPA; the Institute of Electrical and Electronics Engineers (IEEE), where he serves on the Power Systems Grounding Committee ("Green Book"); the International Conference of Building Officials (ICBO); Southern Building Code Congress International (SBCCI); and Building Officials and Code Administrators International (BOCA). Mr. Loyd is currently licensed as a master contractor/ electrician in Arkansas (lic. #1725) and Idaho (lic. #2077) and is an NBEE certified master electrician.

Mr. Loyd served as the chief electrical inspector and administrator for the states of Idaho and Arkansas. He has served as chairman of NFPA 79 Electrical Standard for Industrial Machinery, as a member of the Underwriters Laboratories Advisory Electrical Council, as chairman of the Educational Testing Service (ETS) multistate electrical licensing advisory board, and as a master electrician and electrical contractor. He has been accredited to teach licensing certification courses in Florida, Idaho, Ohio, Oregon, and Wyoming and has taught basic electricity and *NEC®* classes for Boise State University (Idaho).

CHAPTER 1

Codes and Standards in Wiring and Building Design

Purpose

- To provide information for classifying hazardous locations
- To assist those using this book in interpreting the *National Electrical Code*® (*NEC*®) for the purpose of making installations
- To provide information that will assist in minimizing the fire hazards encountered in classified locations
- To provide the reader with information on the various types of electrical equipment and their design used in hazardous locations
- To provide references to sources where more detailed information can be obtained

Hazardous locations in North America can be classified and installed using two distinctly different methods or concepts. Over time, most installers will encounter both of these. Each method takes a totally different approach to accomplish essentially the same levels of protection. The decision on which method is used is up to those making the classification and selecting the equipment. Existing facilities can be reclassified to allow different methods to be used, or some facilities may utilize both methods; however, they cannot be mixed. Both concepts are covered in this manual.

Safety is paramount in an electrical installation where hazardous elements are or may be present. The likelihood that these elements are present and the purpose of the installation must be the prime considerations in designing an electrical system. Over the past 100 years, various codes and standards have been developed to provide guidelines and regulations necessary to ensure safe installations. These guidelines govern electricity from its development (generation) and transmission, including associated equipment, the utilization equipment supplied, and the maintenance and repair of the electrical system after it is installed. Primary in designing a hazardous wiring system, whether it be large industrial facility or a very small hazardous material storage closet, is an understanding of the part played by the codes and standards governing each condition. Codes govern what one can and cannot do, and standards govern safety and technical aspects of the products used. Standards are also a means of ensuring that a variety of parts needed in an electrical system fit together, even when produced by multiple manufacturers. The electrical code is the most important standard for making the electrical installation. However, other standards are necessary to properly classify the area within the facility where hazardous elements are or may be present. In many jurisdictions, fire and building codes can take precedence and govern wiring decisions.

In general, nationally utilized codes and standards are developed by what is known as the consensus process. As we all know, there can be differing views on how or what we install and on the acceptable level of safety to be achieved. For this reason, the consensus method is used, meaning that the pros and the cons are examined, and the language becomes that on which the majority agree.

Each organization has its own basic rules and regulations, but the primary goal is to ensure that the code or standard under development or revision has widespread distribution and opportunity for review by users, producers, and regulators. The code or standard may carry a dual designation; for example, ANSI/UL means that Underwriters Laboratories developed the standard but that public review was through the American National Standards Institute.

National Fire Protection Association Standards

The National Fire Protection Association (NFPA) has acted as the sponsor of the *NEC*® as well as many other safety standards for nearly 100 years. Each year many standards are revised and new standards are developed to meet the needs of our changing world.

NEC®

The most widely used electrical code in the world is the *NEC*®, whose official designation is ANSI/NFPA 70. The *NEC*® was first developed about 100 years ago by interested industry and governmental authorities to provide a standard for the essentially safe use of electricity. It is now revised and updated every three years. Each new edition supersedes all previous editions. It was first developed to provide the regulations necessary for the practical safeguarding of persons and property from the hazards arising from the use of electricity. There are about 4,000 proposals to amend the code each three-year period. These proposals for change are primarily to introduce new products and to clarify the existing text. The *NEC*® is developed by twenty different code-making panels (CMPs) composed entirely of volunteers. These volunteers come from contractors, installers, manufacturers, testing laboratories, inspection agencies, engineering users, governmental agencies, and the electrical utilities.

Suggestions for the content come from a wide array of sources, including individuals just like you. Anyone can submit a proposal or make a comment on a proposal that has been submitted by another individual. For more specifics on the *NEC*® process, you can contact the National Fire Protection Association, Batterymarch Park, Quincy, MA 02269. A free booklet, "The NFPA Standards Making System," is available on request.

The *NEC*® is purely advisory as far as NFPA and ANSI are concerned but is offered for use in law and regulations in the interest of life and property protection. The name "The National Electrical Code" might lead one to believe that this document is developed by the federal government. This is not so; the *NEC*® recognizes the use of products only in a particular way. It approves nothing and has no legal standing until it has been adopted by the *authority having jurisdiction* (see *NEC*® definition Section 90-4), usually a government entity (e.g., federal, state, or city). Therefore, it is necessary to first check with the local electrical inspection department to see which edition of the *NEC*® has been adopted, if any. Compliance with the *NEC*®, coupled with proper maintenance, should result in an installation essentially free from hazard but not necessarily efficient, convenient, or adequate for good service or future expansion of electrical use. To meet the specific requirements of commercial and industrial installations, architects and engineers generally specify beyond the minimum requirements of the *NEC®*, particularly in hazardous installations where factors specific to each installation justify the use of methods and materials that will provide greater protection and an additional level of safety. The *NEC®* is not an instruction manual for untrained persons, nor is it a design specification. However, it does offer design guidelines.

NEC® Section 90-2 defines the scope of the *NEC®*. It is intended to cover all electrical conductors and equipment within or on public and private buildings or other structures, including mobile homes, recreational vehicles, and floating buildings, and other premises, such as yards, carnivals, parking and other lots, and industrial substations. The *NEC®* also applies to installations of conductors and equipment that connect to the supply of electricity, other installations of outside conductors on the premises, and installations of optical fiber cables. The *NEC®* is not intended to cover installations in ships, watercraft, railway rolling stock, aircraft, automotive vehicles, underground mines, and surface mobile mining equipment. It also is not intended to cover installations governed by the utilities, such as communication equipment or transmission, generation, and distribution installations, whether in buildings or on right-of-ways. **Note:** For the complete laundry list of exemptions and coverage, see 1999 *NEC®* Section 90-2.

Using the NEC®

Mandatory rules in the *NEC®* are characterized by the word "shall," such as "shall be permitted" or "shall not be permitted." Explanatory information and references are in the form of "Fine Print Notes," which are identified as (FPN). All tables and footnotes are a part of the mandatory language. Material identified by the superscript letter "x" includes text extracted from other NFPA documents. A complete list of all NFPA documents referenced can be found in the appendix to the *NEC®*. New revisions to the *NEC®* are identified by (a) a vertical line placed in the margin where new text has been inserted and (b) where a bullet (an enlarged black dot) placed in the margin where text has been deleted. **Warning:** Where text has been relocated to another section or article, there is no indication given to alert the user that the requirements are still in the *NEC®*.

To use the *NEC®*, one must first have a thorough understanding of Article 90, "The Introduction"; Article 100, "Definitions"; Article 110, "Requirements for Electrical Installations"; and Article 300, "Wiring Methods." The next step is to become familiar with Article 230 on services, Article 240 on overcurrent protection, and Article 250 on grounding. The rest of the code book can be referred to on an as-needed basis for a particular type of installation.

It is important to remember that one portion of the *NEC®* can override another, sometimes without reference. Therefore, it is imperative to look beyond general wiring methods to the specific type of installation. Some examples are Section 300-22 (on plenums and environmental air spaces), Article 517 (on health care facilities), and Article 518 (on places of assembly). **Note:** Other codes and standards may take precedence over the *NEC®*. The NFPA 101 Life Safety Code, applicable building codes, and fire codes may be more restrictive than this article. All these require specific types of wiring methods under certain conditions.

The installer should verify the environment for the installation before selecting the wiring method. Choosing and installing the wrong method before checking the *NEC®* thoroughly may prove very costly when an inspector notifies you that the installation has to come out!

Other NFPA Standards

Other NFPA standards relating to hazardous electrical installations and their safety are as follows:

- NFPA 30 Flammable and Combustible Liquids Code
- NFPA 30A Automotive and Marine Service Station Code
- NFPA 88A Parking Structures
- NFPA 88B Repair Garages
- NFPA 99 Health Care Facilities
- NFPA 101 Life Safety Code
- NFPA 496 Purged and Pressurized Enclosures for Electrical Equipment
- NFPA 497 Recommended Practice for the Classification of Flammable Liquids, Gases, or Vapors and of Hazardous (Classified) Locations for Electrical Installations in a Chemical Process Areas (formerly NFPA 497A and 497M).
- NFPA 499 Recommended Practice for the Classification of Combustible Dusts and of Hazardous (Classified) Locations for Electrical Installations in a Chemical Process Areas (Formerly NFPA 497B).

Note: These are only examples of the many NFPA standards that are applicable to specific conditions and installations. You might need to refer to many other publications in the analysis of the conditions and areas to be classified. Many of these publications can be referenced in the appendices of the various NFPA standards: the Occupational Safety and Health Act of 1970 (OSHA), the American Petroleum Institute (API), American Society of Testing and Materials (ASTM) publications, the U.S. Bureau of Mines, and the National Academy of Sciences. Designers and installers should review the applicable standards before designing or making hazardous installations.

American National Standards Institute

The American National Standards Institute (ANSI) is an umbrella organization for the consensus code-making process. ANSI has some of its own volunteer committees that develop standards that are published and sold. In addition, ANSI is an organization that coordinates the efforts of several other organizations that develop codes and standards for the electrical industry; including the National Fire Protection Association (NFPA), the Institute of Electrical and Electronic Engineers (IEEE), and Underwriters Laboratories (UL). After a standard has gone through the ANSI process and received public review, the standard then becomes an "American National Standard." The technical information found in the purely ANSI standards may be identical or similar to that found in the dual designated UL, *NEC*®, and IEEE standards, which are identified as ANSI/UL. The designations ANSI/NFPA 70 (the *NEC*®) or ANSI/IEEE/ANSI/NEMA indicate that the document has been circulated for comment through the ANSI system.

NEMA and IEC Standards

The National Electrical Manufacturers Association (NEMA) also develops standards for some products. These standards are processed through the ANSI mechanism. NEMA standards are developed by manufacturers' representatives who belong to various NEMA sections (industry groups). They write from intimate knowledge of the products they produce. These standards are to provide product consistency, information, and safety.

Some major NEMA standards deal with switchgear and overcurrent protection devices. The major NEMA raceway standard is RN-1, which covers plastic-coated tubular steel; there is no UL or ANSI standard counterpart for this product. The most common of the

NEMA standards are the NEMA enclosure types. Electricians commonly order panel-boards, safety switches, and junction boxes by the NEMA enclosure type, such as a NEMA 3R for raintight or a NEMA 1 for an indoor, dry location. Enclosures for hazardous locations include the NEMA 7 and the NEMA 10. When these enclosures are properly installed and maintained, they are designed to contain an internal explosion without causing external hazard. NEMA Type 7 enclosures are designed for indoor use in locations classified as Class I, Groups A, B, C, and D. NEMA Type 8 enclosures are designed to prevent combustion through the use of oil-immersed equipment. Type 8 is suitable for use in areas classified Class 1, Groups A, B, C, and D, indoor or outdoor. NEMA Type 9 enclosures are designed to prevent the ignition of combustible dust and are suitable for locations classified Class II, Groups E, F, and G. NEMA Type 10 are nonventilated types capable of meeting 30 C.F.R. Part 18 of the Mine Safety and Health Administration (MSHA).

NEMA also coordinates much of the input to documents designed by the International ElectroTechnical Commission (IEC). This organization is based in Europe and is engaged in writing electrical standards for use around the world. The United States is only one of many countries providing input to these standards and has only one vote. Little by little, IEC standards are making their way into this country. However, they can have some modifications and generally are issued as a UL version of the IEC standard. IEC is primarily mentioned here so you will know when the term shows up in a specification or as a label on a piece of equipment. It is suggested that you read the magazine articles as they appear on IEC- and EC364-related subjects to gain knowledge of overseas activity that will affect the electrical industry in this country for years to come.

Federal Government Standards

The federal government had its own standards for most products until the late 1970s. In the interest of economics, there was a move toward adopting industry standards and canceling the special government specifications, which not only took much time to prepare but also created excessive purchase costs because special manufacturing was unnecessarily required. Because many designer specifications still contain outdated references to federal specifications, the old number will be referenced in this book for correlation. For the most part, ANSI and UL standards have now been adopted by the federal government for the wiring systems discussed here.

Local Codes and Requirements

Although most municipalities, counties, and states adopt the *NEC®*, they may not have adopted the latest edition, and in some cases the authority having jurisdiction (AHJ) might be several editions of the *NEC®* behind. Most jurisdictions make amendments to the *NEC®* or add local requirements. These may be based on environmental conditions, fire safety concerns, or other local experience. An example of one quite common local amendment is that all commercial buildings be wired in raceways. The *NEC®* generally does not differentiate between wiring methods in residential, commercial, or industrial installations; however, many local jurisdictions do.

Metal wiring methods, especially in fire zones, is another common amendment. Some major cities have developed their own electrical code (e.g., Los Angeles, New York, Chicago, and several metropolitan areas in Florida). In addition to the *NEC®* and all local amendments, the designer and installer must comply with the local electrical utility rules. Most utilities have specific requirements for installing the service to the structure. There have been many unhappy designers and installers who have learned about special jurisdictional requirements after making the installation, thereby incurring costly and time-consuming corrections at their own expense.

As an example, when I was administrator and chief electrical inspector for the State of Idaho, I encountered a situation where the health department and EPA required many homeowners desiring to build near lakes, rivers, and other waterways to locate their septic systems a prescribed distance away from the water's edge. This was often several hundred feet. To accomplish this generally required the installation of a small sewage holding tank with a sewage pump to move the sewage away from the water to a safe distance where the septic system could be safely located. We recognized that there was a possibility of the accumulation of methane gas in this small pumping station. Because of that possibility, it had to be classified as a Class I, Division 1 location. However, to require all homeowners to build or purchase a tank that would meet the requirements of a Class I, Division 1 area was very expensive. Therefore, with the cooperation of the Electrical Division of the Idaho Department of Labor, a tank was designed and built that would limit the hazardous area to a small area where only the conductors passed through and contained no splices or equipment. These items were located to provide the degree of safety required at the lowest possible cost.

The pump motor was submersed, and the low-level float switch was located above the top of the pump motor. An additional redundant emergency kill float switch was also located above the top of the motor but just below the low-level switch. By doing this, we permitted a standard pump (not a listed explosion-proof pump). The area above the water level within the tank was classified as a Class I, Division 1 location. This area contained only the unbroken cord from the motor. The cord then entered rigid metal conduit or intermediate metal conduit in compliance with Section 501-4(a). An explosion-proof seal was required within 18 inches and splices were made in an explosion-proof box. A general purpose controller was used to control the system.

It is often possible to work together to provide the level of safety necessary by limiting the hazardous area to a minimum and relocating arc-producing equipment out of those areas, thereby eliminating the need for expensive equipment approved for classified locations (see Figure 1–1).

Insurance Requirements

Many insurance companies have requirements that will affect the insurability and rates of the structure. Those requirements should be determined prior to completing the design stage. The degree of risk to the insurer will be an important determining factor. It may be possible to make design changes that will greatly improve the insurability and rates. The changes necessary to achieve these desired benefits can involve architectural and ventilation changes; therefore, all interested parties should be involved. In some cases, the wiring methods used and the electrical wiring design only may affect such improvement.

Nationally Recognized Testing Laboratories (NRTL)

Underwriters Laboratories (UL) has long been the major product testing laboratory in the electrical industry. In addition to testing, UL is a longtime developer of product standards. By producing to these standards and contracting for follow-up service after undergoing a listing procedure, a manufacturer is authorized to apply the UL label or to mark the product. In the last decade, numerous other electrical testing laboratories have come on the scene and are being officially recognized. Underwriters Laboratories is not the only testing laboratory evaluating electrical products any more. There are many testing laboratories operating today. Some jurisdictions evaluate and approve laboratories; others accept them on the basis of reputation. The Occupational Safety and Health Administration (OSHA) is now evaluating and approving testing laboratories. It is the duty of the entity responsible for specifying the materials to verify that the product has been evaluated by a testing

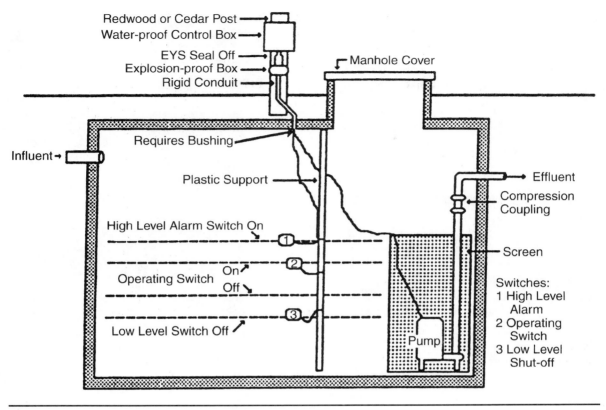

Figure 1–1   Example of a sewage lift pumping station that is acceptable by the AHJ.

Figure 1–2 Examples of the trademarks of several well-known testing laboratories.

laboratory acceptable to the authority having jurisdiction where the installation is being made. It is the installer's responsibility to ensure that the product is installed in accordance with the manufacturer's instruction and listing (*NEC*® Section 110-3[b]).

On most products you will find a "Listing Mark" and/or a "Label." Some AHJs will require that you install only listed products in their jurisdiction. Some of the better-known testing agencies are Metropolitan Electrical Testing (MET), ETL Testing Laboratories (ETL), and Factory Mutual (FM). Examples of testing agency trademarks are shown in Figure 1–2.

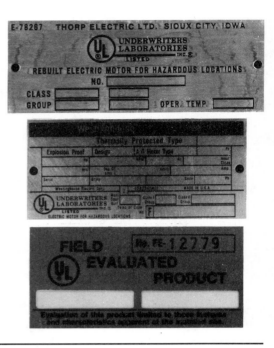

Figure 1–3 Examples of different labels as they appear on equipment. (Courtesy of Underwriters Laboratories.)

These testing laboratories generally do not write standards. The testing they perform is performed to UL-, ANSI-, or ASTM-established standards. All except FM have not yet ventured into writing standards.

NEC® Section 110-3(b) requires that you install products in accordance with the manufacturer's instructions, which are part of the listing, and in accordance with the *NEC®*. The instructions may or may not be included with the product (see Figure 1–3). The testing laboratory evaluates products according to a standard that is designed to ensure that all manufacturers of a specific item meet a minimum requirement that has been developed on the basis of the expected use in accordance with the *NEC®* (see Figure 1–4). Many electrical system failures, accidents, and fires originate from products that were installed incorrectly or were not the correct product for the application, which is a violation of *NEC®* Section 110-3(b). It is interesting to note that although most general purpose motors are *not* UL listed, most motors designed for use in hazardous (classified) locations are UL listed. The UL "Green Book" (*Electrical Constructions Material Directory*), UL "White Book" (*General Information for Electrical Construction, Hazardous Location, and Electric Heating and Air-Conditioning Equipment*), and the UL "Red Book" (*Hazardous Location Equipment*) will provide valuable information that can help avoid the misapplication of materials. These are the only documents that provide information regarding limitations and specific requirements for a product listing and should be a part of every designer's and installer's library. These and other UL directories and product standards can be obtained from Underwriter's Laboratories Inc., 333 Pfingsten Road, Northbrook, IL 60062.

IDENTIFICATION OF UL LISTED PRODUCTS THAT ARE ALSO UL CLASSIFIED IN ACCORDANCE WITH INTERNATIONAL PUBLICATIONS

Underwriters Laboratories Inc. (UL) provides a service for the Classification of products that not only meet the appropriate requirements of UL but also have been determined to meet appropriate requirements of the applicable international publication(s). For those products which comply with both the UL requirements and those of an international publication(s), the traditional UL Listing Mark and a UL Classification Marking, as described below, may appear on the product as a combination Listing and Classification Marking.

The combination of the UL Listing Mark and Classification Marking may appear as authorized by Underwriters Laboratories Inc.

LISTED
(Product Identity)
(Control Number)

ALSO CLASSIFIED BY UNDERWRITERS LABORATORIES INC.® IN ACCORDANCE WITH IEC PUBLICATION _____

LISTED
(Product Identity)
(Control Number)

ALSO CLASSIFIED BY UNDERWRITERS LABORATORIES INC.® IN ACCORDANCE WITH IEC PUBLICATION _____

UND. LAB. INC. ®   LISTED
(Product Identity)
(Control Number)

ALSO CLASSIFIED BY UNDERWRITERS LABORATORIES INC.® IN ACCORDANCE WITH IEC PUBLICATION _____

Underwriters Lab. Inc. ®   LISTED
(Product Identity)
(Control Number)

ALSO CLASSIFIED BY UNDERWRITERS LABORATORIES INC.® IN ACCORDANCE WITH IEC PUBLICATION _____

(Product Identity)
CLASSIFIED BY UNDERWRITERS LABORATORIES INC.® IN ACCORDANCE WITH IEC PUBLICATION _____
(Control Number)
OR
(Product Identity)
CLASSIFIED BY UNDERWRITERS LABORATORIES INC.® IN ACCORDANCE WITH IEC PUBLICATION _____
(Control Number)

LOOK FOR THE CLASSIFICATION MARKING

(Product Identity)
(Control Number)

ALSO CLASSIFIED BY UNDERWRITERS LABORATORIES INC.® IN ACCORDANCE WITH IEC PUBLICATION _____

LISTED
(Product Identity)
(Control Number)

ALSO CLASSIFIED BY UNDERWRITERS LABORATORIES INC.® IN ACCORDANCE WITH IEC PUBLICATION _____

UNDERWRITERS LABORATORIES INC. ®   (Product Identity)
LISTED   (Control Number)

ALSO CLASSIFIED BY UNDERWRITERS LABORATORIES INC.® IN ACCORDANCE WITH IEC PUBLICATION _____

LOOK FOR THE COMBINATION LISTING MARK AND CLASSIFICATION MARKING

IDENTIFICATION OF PRODUCTS CLASSIFIED TO INTERNATIONAL PUBLICATIONS ONLY

Underwriters Laboratories Inc. (UL) provides a service for the classification of products that have been determined to meet the appropriate requirements of the applicable international publication(s). For those products which comply with the requirements of an international publication(s), the Classification Marking may appear in various forms as authorized by Underwriters Laboratories Inc. Typical forms which may be authorized are shown below, and include one of the forms illustrated, the word "Classified", and a control number assigned by UL. The product name as indicated in this Directory under each of the product categories is generally included as part of the Classification Marking text, but may be omitted when in UL's opinion, the use of the name is superfluous and the Classification Marking is directly and permanently applied to the product by stamping, molding, ink-stamping, silk screening or similar processes.

Separable Classification Markings (not part of a name plate and in the form of decals, stickers or labels) will always include the four elements: UL's name and/or symbol, the word "Classified", the product category name, and a control number.

The complete Classification Marking will appear on the smallest unit container in which the product is packaged when the product is of such a size that the complete Classification Marking cannot be applied to the product or when the product size, shape, material or surface texture makes it impossible to apply any legible marking to the product. When the complete Classification Marking cannot be applied to the product, no reference to Underwriters Laboratories Inc. on the product is permitted.

Figure 1–4 Identification marks for different types of listings and classifications. (This material extracted from Underwriters Laboratories, Inc., *General Information for Electrical Construction, Hazardous Location and Electrical Heating and Air-Condition Equipment Directory.* Courtesy of Underwriters Laboratories.)

Occupational Safety and Health Act of 1970

OSHA began developing and issuing standards intended to protect employees from hazards in the workplace about 1970. In an attempt to eliminate electrical hazards in the workplace, OSHA has continued, and continues today, to develop consensus standards to address specific safety concerns associated with design, installation, maintenance, and repair of electrical systems and components. General safety standards have been developed that cover construction sites and the maintenance and repair of electrical powered machines and equipment. Every electrical system, new or existing, must satisfy these requirements. They are fully retroactive, without reference to time. Any violations of these rules constitute OSHA violations and are cause for citations.

All electrical systems must be in complete compliance with all the rules, or they are in violation of the OSHA law and must be altered or retrofitted to produce full compliance with every one of the OSHA rules. Electrical installations made after March 15, 1972, are covered by the OSHA Act of 1970. All were required to comply with the rules in Sections 1910.302 through 1910.308, with eleven specific exceptions. Additional requirements were added effective April 16, 1981. All the existing electrical installations were required to comply with the rules in Sections 1910.302 through 1910.308, without exception.

Today, the new design and safety standards for electrical systems are found in Subpart S, Part 1910 of Title 29 of the *Code of Federal Regulations*. The entire document is divided into major sections and numerically designed as follows:

- 1910.146 Permit Required for Confined Spaces for General Industry. Effective April 1, 1993
- 1910.147 Control of Hazardous Energy (Lockout/Tagout)
- 1910.269 Electric Power Generation, Transmission, and Distribution; Electrical Protective Equipment
- 1910.302 Electrical Utilization Systems
- 1910.303 General Requirements
- 1910.304 Wiring Design and Protection
- 1910.305 Wiring Methods, Components, and Equipment for General Use
- 1910.306 Specific Purpose Equipment and Installations
- 1910.307 Hazardous (Classified) Locations
- 1910.308 Special Systems
- 1910.399 Definitions Applicable to This Subpart

The newly adopted 29 CFR 1910.146 sets access to restricted or limited occupancy space areas not designed for continuous occupancy and are hazardous or may be hazardous (e.g., grain silos). These new rules require the employer to set up a sequence of operations, such as the following:

- Identify hazard(s).
- Identify the presence and availability of oxygen.
- Set up a system to identify each employee entering the work assignments.
- Prevent unauthorized entry.
- Provide necessary protective equipment.
- Provide needed protective clothing.
- Provide at least one worker outside the area to summon help if needed.
- Provide briefing and debriefing to inform all employees and contractors of these procedures.

The employer is responsible for naming supervisors or designated persons responsible for carrying out each of these tasks.

The OSHA requirements state that all electric products, conductors, and equipment must be approved (listed and/or labeled by a nationally approved testing laboratory, or they must be replaced with products that are acceptable by OSHA definition, and they must be installed in accordance with the manufacturer's instructions included in the listing or labeling for every installation requiring the use of approved equipment. The rule clearly states that any and every product listed in the UL *Electrical Construction Material Directory* or in any other listing directory must be used exactly as described in the application data given in the listing in the book. OSHA is presently evaluating and approving testing laboratories. It is necessary to consult the Federal Register to determine which nationally approved testing laboratories are acceptable to the OSHA guidelines.

OSHA has announced the availability of an OSHA fax service. Callers can dial 1 (900) 555-3400 to obtain information on various OSHA regulations. The system offers several hundred documents at a cost of $1.50 per minute, billed on your telephone bill. A touch-tone telephone and fax machine are required to utilize this service. Callers can request a maximum of two documents per call.

The database will contain agency news releases, fact sheets, publication listings, OSHA office listings, inspection statistics, and other information. Most of this information can also be obtained by mail through OSHA Publications, Room N3101, 200 Constitution Ave. NW, Washington, DC 21210.

CHAPTER

2

Basic Design Factors

Several factors must be considered together in order to achieve a well-designed electrical system for a building or structure (see Figure 2–1). Consideration must be given to the safety requirements. This is particularly important for facilities where hazardous substances are handled or stored. The safety considerations include the wiring methods and architectural considerations, such as means of egress and area separation. The capacity of the electrical service must be considered, as does voltage regulation. The service should be large enough to handle future expansion, and the voltage should be selected accordingly. Consideration should also be given to accessibility and flexibility. These factors are applicable to the complete distribution or utilization system or part thereof. The ultimate goal is to provide an electrical system that will be efficient to operate, meet initial as well as future requirements, and lend itself to easy maintenance and economical alterations.

Before drawing the plans, a few things must be known so that a system can be designed that will provide "adequacy" (Section 90-1[b]). The *NEC®* states that the minimum requirements of the *NEC®* will provide an installation that is safe and free from electrical hazards, but the installation may not provide all the convenience we would like nor provide provisions for future expansion.

Safety

The initial classification prior to the designer developing the plans and specifications provides a foundation on which a safe electrical installation can be designed and installed. The initial design is based on the first four chapters of the *NEC®*. The hazardous (classified) areas of the electrical installation require far greater considerations (see figure 2–2). It is necessary to include the requirements of chapters one through four as applicable. The wiring methods and equipment must be selected based on the requirements in the appropriate article(s) in Chapter 5. Other NFPA standards will need to be used for area classification and to provide guidance in making the final decision on the system design and the selection of equipment and methods. Close cooperation with the authority having jurisdiction, along with others such as the fire marshal and insurance company officials, is needed. The input of a knowledgeable designing engineer is essential to ensure that the electrical system incorporates safeguards for the specific conditions. Consideration must be

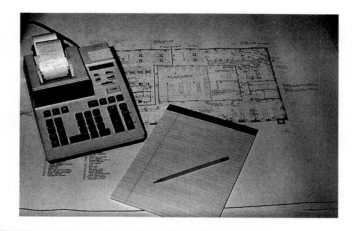

Figure 2–1 Basic design factors must be known before beginning.

What Are Hazardous (Classified) Locations?

Locations are classified on the basis of the properties of the flammable vapors, liquids or gases, or combustible dusts of fibers that might be present and the likelihood that a flammable or combustible concentration or quantity is present.

Each room, section, or area must be individually classified by indicating:

_____ Class	I	II	III
_____ Division	1 or 2	1 or 2	1 or 2
_____ Group	A–D	E–G	—

Article 505 is optional for Class I only.

_____ Zone	0, 1, or 2	—	—
_____ Group	IIC, IIB, or IIA	—	—

Figure 2–2 Careful consideration must be given and all classifications made prior to completing the design phase.

---

given to location, occupancy, use, and the possibilities of future expansion. The electrical design should include initial as well as future power capabilities.

NEC® Article 305 and OSHA offer guidelines and requirements for safe, temporary electrical systems and safe working conditions during the installation (construction) stage.

Capacity

It is not uncommon for an electrical system to provide continual service for fifty years or more. Therefore, the designed capacity of a new installation must receive comprehensive consideration and will often require close coordination with the serving utility, the structure owner, and the designer. Although oversizing a new system will provide good future

Figure 2–3   It makes sense to consider expansion when specifying the distribution system. (Courtesy of Square D Company.)

expansion capabilities at a relatively low cost, it may not be usable when needed if not closely coordinated with all interested parties (see Figure 2–3).

Special attention should be given to the anticipated needs as well as the future contingency of an electrical service. Consideration should be given to air-conditioning and heating requirements, elevators, data processing and computers, indoor and outdoor lighting, and processing equipment. Future automation and state-of-art upgrading in industrial, commercial, institutional, and residential installations must be considered. It is recommended that service equipment, distribution equipment (switchboards and panelboards), feeders, raceways and conductors, junction boxes, and cabinets be specified at least one size larger than the calculated electrical load would require (for sixing raceways, see Figures 2–4 and 2–5). Branch-circuit panelboards should contain no less than 10%, preferably 25%, space circuit breaker or fuse spaces for expansion capabilities. Working space (physical room) required by *NEC*® Sections 110-26 and 110-32 should also provide additional room for future expansion. Spare raceway capacity simplifies rewiring and is a significant factor in minimizing the cost when such work is undertaken. **Note:** The additional expansion capabilities recommended in the section represent good design, although the *NEC*® does not require this.

The information in Figures 2–4 and 2–5 has been formulated from the newly developed grounding and electromagnetic interference (GEMI) analysis software developed by Dr. A.P. "Sakis" Meliopolous at the School of Electrical and Electronic Engineering at the Georgia Tech Research Institute. These tests coincide with extensive changes to the *NEC*® that added exact dimension tables for each type of wire and raceway to Chapter 9, Tables 4 and 5, and new Appendix C for fill capacity. An additional change added a note to warn the user that the minimum-size-equipment grounding conductor listed in the table might

Examples of Maximum-Length Equipment Grounding Conductor (Steel EMT, IMC, GRC and Copper, Copperclad, or Aluminum Wire Computed as a Safe Return Fault Path to Overcurrent Device Based on 1994 Georgia Tech Software (GEMI 2.3) with an Arc Voltage of 40 and 4 IP at 25°C Ambient

120 Volts to Ground

Overcurrent Device Rating, Amperes (75°C)	400% (4IP) Overcurrent Device Rating, Amperes	Circuit Conductor Size (AWG-kcmil) Copper or Aluminum	EMT, IMC GRC Trade Size	(1) Equipment Grounding Conductor Size, Copper or Aluminum	Length of EMT Run Computed Maximum (in Feet)	Length of IMC Run Computed Maximum (in Feet)	Length of GRC Run Computed Maximum (in Feet)	(1) Copper Grounding Conductor Max Run (in Feet)	(1) Aluminum or Copperclad Grounding Conductor Max Run (in Feet)
20	80	12	1/2	—	395	398	384	—	—
20	80	12	—	12	—	—	—	300	—
20	80	10 AL	—	10 AL	—	—	—	—	293
30	120	10	1/2	—	—	383	—	—	—
30	120	10	3/4	—	404	399	386	—	—
30	120	10	—	10	—	—	—	319	—
30	120	8 AL	—	8 AL	—	—	—	—	310
40	160	8	3/4	—	—	414	—	—	—
40	160	8	1	—	447	431	418	—	—
40	160	8	—	10	—	—	—	294	—
40	160	8 AL	—	8 AL	—	—	—	—	232
60	240	6	1	—	404	400	382	—	—
60	240	6	—	10	—	—	—	228	—
60	240	4 AL	—	8 AL	—	—	—	—	221
100	400	3	1 1/4	—	402	397	373	—	—
100	400	3	—	8	—	—	—	229	—
100	400	1 AL	—	6 AL	—	—	—	—	222
200	800	3/0	2	—	390	389	363	—	—
200	800	3/0	—	6	—	—	—	201	—
200	800	250 AL	—	4 AL	—	—	—	—	195

(1) Per NEC® Table 250-95.
Applicable to nonmetallic conduit runs.

*NEC® wire fill table permits smaller conduit size.
Note: Software GEMI 2.4 is not limited to above examples.

Figure 2–4 Table showing examples of acceptable equipment grounding conductor lengths using steel conduit and tubing as the equipment grounding conductor on circuits 120 volts to ground.

Examples of Maximum-Length Equipment Grounding Conductor (Steel EMT, IMC, GRC and Copper, Copperclad, or Aluminum Wire Computed as a Safe Return Fault Path to Overcurrent Device Based on 1994 Georgia Tech Software (GEMI 2.4) with an Arc Voltage of 40 and 4 IP at 25°C Ambient

277 Volts to Ground

Overcurrent Device Rating, Amperes (75°C)	400% (4IP) Overcurrent Device Rating, Amperes	Circuit Conductor Size (AWG-kcmil) Copper or Aluminum	EMT, IMC GRC Trade Size	(1) Equipment Grounding Conductor Size, Copper or Aluminum	Length of EMT Run Computed Maximum (in Feet)	Length of IMC Run Computed Maximum (in Feet)	Length of GRC Run Computed Maximum (in Feet)	(1) Copper Grounding Conductor Max Run (in Feet)	(1) Aluminum or Copperclad Grounding Conductor Max Run (in Feet)
20	80	12	1/2	—	1,170	1,179	1,140	—	—
20	80	12	—	12	—	—	—	890	—
20	80	10 AL	—	10 AL	—	—	—	—	870
30	120	10	1/2	—	—	1,135	—	—	—
30	120	10	3/4	—	1,199	1,182	1,143	—	—
30	120	10	—	10	—	—	—	946	—
30	120	8 AL	—	8 AL	—	—	—	—	920
40	160	8	3/4	—	—	1,228	—	—	—
40	160	8	1	—	1,326	1,276	1,239	—	—
40	160	8	—	10	—	—	—	871	—
40	160	8 AL	—	8 AL	—	—	—	—	690
60	240	6	1	—	1,197	1,186	1,131	—	—
60	240	6	—	10	—	—	—	676	—
60	240	4 AL	—	8 AL	—	—	—	—	657
100	400	3	1 1/4	—	1,192	1,176	1,107	—	—
100	400	3	—	8	—	—	—	680	—
100	400	1 AL	—	6 AL	—	—	—	—	659
200	800	3/0	2	—	1,157	1,155	1,077	—	—
200	800	3/0	—	6	—	—	—	598	—
200	800	50 AL	—	4 AL	—	—	—	—	578

(1) Per NEC® Table 250-95.
Applicable to nonmetallic conduit runs.

*NEC® wire fill table permits smaller conduit size.
Note: Software GEMI 2.4 is not limited to above examples.

Figure 2-5 Table showing examples of acceptable equipment grounding conductor lengths using steel conduit and tubing as the equipment grounding conductor on circuits 277 volts to ground.

not be large enough in some cases. Figures 2–4 and 2–5 show several common examples of acceptable circuit lengths and conduit and tubing types. The study also covers the electromagnetic field analysis and the most effective conduit for shielding these fields where they are causing interference or other power quality problems.

Flexibility

A good electrical system provides flexibility. The distribution system should be arranged to accommodate future physical changes in plant or office layout as production lines are changed, equipment shifted, or offices rearranged. The increased popularity of modular office layouts demands an electrical system flexibility to meet quick and economical rearrangement. One or more raceway types will offer the versatility and flexibility needed to meet today's constantly changing world (see [FPN] following the definition of "Raceway" in Article 100 for a list of raceway types).

Steel tubular conduit (Rigid Metal Article 346, IMC Article 345, and EMT Article 348) offers almost unlimited expansion possibilities and is adaptable to all types of building construction and floor plans. In addition, cellular and underfloor raceways offer creative circuit and outlet arrangement. Similarly, surface raceway, busway, and wireways provide additional flexibility in the design of industrial, institutional, and commercial facilities.

Flexibility can be maintained only if adequate additional space is provided in the panelboards and switchboards (see dedicated space requirements *NEC*® Section 110-26f). Additional space should be provided, if possible, to add feeders in the future, or spare conduits can be installed to areas that either are not accessible or where future installation would be labor intensive, such as underfloor, buried in concrete, or in shafts. The importance of flexibility cannot be overemphasized.

Electrical Systems and Building Fires

The prime objectives of the *NEC*® are to provide safety from fire and electrical shock (see Figure 2–6). Many of the Code's requirements are based on the premise that an electrical system designed and installed in accordance with the *NEC*® must not start a fire. In addition, an electrical installation must not contribute to or allow the spread of fire should one

Figure 2–6   Example of fire damage caused by an electrical system failure. (Courtesy of Allied Tube & Conduit.)

originate from other causes. The possible contribution of an electrical system to a building fire is undergoing more and more scrutiny as these systems become more extensive and as more uses are proposed for an increasing variety of electrical raceway materials.

Fire Resistance and Spread of Fire and Smoke

NFPA 101, Fire and Life Safety Code, and the three major model building codes (ICBO, BOCA, and SBCCI), plus a variety of local, state, and city building codes, include many provisions whose purpose is to provide building construction that will not only withstand the effects of fire in a portion of a building but also reduce the likelihood of fire spreading within the building. For extremely large or high structures, building codes usually require fire-resistive construction as well as the use of noncombustible construction materials. Building codes prescribe that for fire-resistive buildings, the columns, beams, girders, walls, and floor constructions have a resistance to fire that is measured by a standard fire test procedure (ASTM E119, Standard Method of Fire Test for Building Construction and Materials).

The fire resistance of a structural assembly is determined from end-point criteria included in this fire test standard. For load-bearing members, such as columns, beams, and girders, maintenance of the load-bearing capability during and after fire exposure is a prime criterion. The temperature rise on the unexposed surface of an assembly is critical for walls and floors aimed at preventing the ignition of combustibles and the spread of fire to the opposite side of such wall or floor assemblies. It is extremely important in fire-resistive construction that all building components adhere to approved design of fire-rated construction. Furthermore, during construction every effort should be made to avoid damage to the fire-resistive integrity of the construction by removing fire-protective coatings or coverings or by piercing through entire fire-rated assemblies.

CHAPTER

3

Hazardous Locations

The engineers and contractors of today handle the design and installation of electrical systems with a high level of confidence. The subject of hazardous (classified) locations, however, is still viewed with uncertainty by many people in the industry.

Actually, hazardous location installations are more exacting than difficult or mysterious. Article 500 of the *NEC*® defines various types of hazardous locations, and Articles 501 through 517 provide hazardous performance and installation requirements.

There are many rules to guide you; however, assurance of safety starts with a good understanding of two basics: proper classification of locations and selection of correct equipment.

When an electrical system, or any portion of it, is to be installed in an area containing a hazardous atmosphere, protection against possible explosions becomes a major design factor. All the basic design considerations still apply to determine conductor sizes and general layout of the overall system. In addition, the portion of the system in or passing through an area where flammable gases, combustible dust, or ignitable fibers may exist must have equipment enclosures especially designed to prevent electrical sparks and arcs from igniting the surrounding atmospheres and causing serious fires and/or explosions. For this portion of the design, the electrical engineer must be careful to select equipment specifically approved for the conditions encountered (see Figures 3–1, 3–2, and 3–3.) **Warning:** *NEC*® Section 90-3 states that Chapter 5 only modifies and augments Chapters 1 through 4. Corrosion, capacity, suitability, and service are not considered. Therefore, it is the responsibility of the designer to take these factors into account, along with the hazardous location requirements, which are covered in Chapter 5.

Hazardous (Classified) Areas

Hazardous (classified) locations are classified on the basis of the properties of flammable gases, vapors, liquids, combustible dusts, fibers, and flyings which may be present and the likelihood that a concentration of flammable or combustible materials is present in sufficient quantity. Areas or sections of each building or room can be classified individually by indicating the class, the division, and the group present.

Figure 3–1 Electrical panelboard suitable for Class II and III, Groups E, F, and G. (Courtesy of Crouse-Hinds, Division of Cooper Industries.)

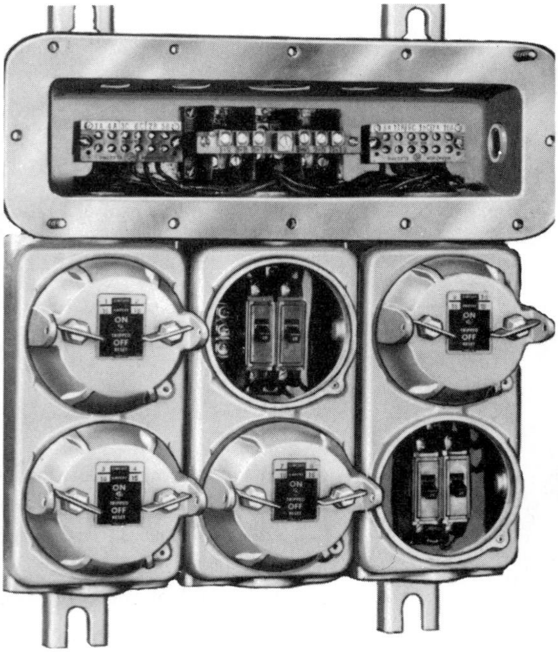

Figure 3–2 Electrical panelboard suitable for Class I, II, and III, Groups C, D, E, F, and G. (Courtesy of Crouse-Hinds, Division of Cooper Industries.)

The *NEC®* describes three classifications of hazardous locations where an unconfined spark could result in explosion or fire (see Figure 3–4). Each of these classifications identifies conditions that may exist under normal operations, under occasional but frequent conditions resulting from maintenance operations, or under conditions of accidental breakdown that may cause the release of flammable or readily combustible materials, *NEC®* Sections 500 through 505, 510, 511 and Sections 513 through 517 contain rules covering the type of wiring and apparatus to be installed in hazardous locations.

The actual degree of hazard (classification) is frequently evaluated by local inspection and fire department authorities, and for that reason they should be consulted during the design stage of the hazardous area wiring system.

"Big Blasts"

Some of this nation's major accidents that have resulted in severe damage to property and loss of life have occurred in the chemical petroleum industries, food processing industries, and other locations containing hazardous areas. These facilities have qualified maintenance

National Electrical Code
Classes, Divisions (Zones), and Groups

Class I

Division 1		*Division 2*
Group A		Group A
Group B		Group B
Group C		Group C

Zone 0	*Zone 1*	*Zone 2*
Group IIC	Group IIC	Group IIC
Group IIB	Group IIB	Group IIB
Group IIA	Group IIA	Group IIA

Class II

Division 1		*Division 2*
Group E		Group E
Group F		Group F
Group G		Group G

Class III

Division 1	*Division 2*

Class III is not divided by groups.

Figure 3–3

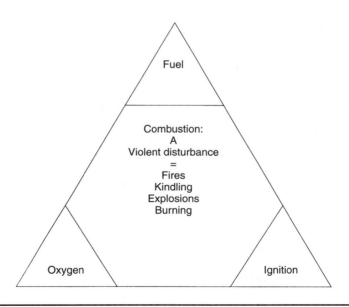

Figure 3–4   These three ingredients combined are needed to have a fire.

personnel and engineering, and they have taken precautions to avoid such accidents. However, the following examples emphasize the dangers in dealing with hazardous classified materials:

- April 8, 1992: A natural gas pipeline explodes near Brenham, Texas, killing one.
- September 3, 1991: A chicken-processing plant of Imperial Food Products Co. near Hamlet, North Carolina, catches fire, killing twenty-five.
- June 17, 1991: Tenneco Inc.'s phosphorus chemicals plant in Charleston, South Carolina, explodes, killing six.
- May 1, 1992: IMC Fertilizer Angus Chemical Co. nitro-paraffins plant in Sterling, Louisiana, explodes, killing eight.
- March 12, 1991: Union Carbide petrochemical plant in Seadrift, Texas, explodes, killing one.
- March 3, 1991: Citgo Petroleum Corporation petroleum refinery near Lake Charles, Louisiana, explodes, killing six.
- July 19, 1990: BASF Corporation in Cincinnati, Ohio, chemical plant explodes, killing two.
- July 5, 1990: Arco Chemical Co. petrochemical plant in Channelview, Texas, explodes, killing seventeen.
- October 23, 1989: Phillips Petroleum Co. plastics plant in Houston, Texas, explodes, killing twenty-three.
- December 24, 1989: Exxon Company petroleum refinery in Baton Rouge, Louisiana, explodes, killing two.
- May 4, 1988: Pacific Engineering and Production Company ammonium prechlorate plant in Henderson, Nevada, explodes, killing two.
- November 5, 1985: Warren Petroleum plant in Warren Belvieu, Texas, explodes, killing two.
- July 1984: Unocal Corporation petroleum refinery in Lemont, Illinois, explodes, killing seventeen.

NEC® Classifications

The *NEC®* divides hazardous locations into three classes, plus divisions and groups that are defined as follows. These are based on the degree of hazard and the likelihood of this hazard being present. The general information regarding these areas is covered in Article 500. Each class is specifically covered by a separate article (e.g., Class I, Article 501 or Article 505; Class II, Article 502; and Class III, Article 503). Some common hazardous installations are specifically covered by an article (e.g., Article 514, Gasoline Dispensing, Service Stations).

Important fact: Article 514 covers specific requirements for gasoline-dispensing service stations; one must also apply pertinent sections of Articles 502 and 500 and Chapters 1 through 4 to make the installations.

Classes

The NFPA standards divide the hazardous properties into three classes so that the installation containing one or more of those properties can be maintained and operated safely using the equipment and wiring methods that will operate safely with a minimum of overdesign.

Class I

This classification applies to locations where the hazard of fire or explosion may exist because of the presence of flammable liquids, gases, or vapors in quantities sufficient to produce an explosive or ignitable environment. **Warning:** Class III locations are not necessarily less hazardous than Class I locations. Major life and property losses have occurred in Class I, Class II, and Class III locations.

Class II

This classification applies to locations where the hazard of fire or explosion may exist because of the presence of combustible dust.

Class III

This classification applies to locations where the hazard of fire or explosion may exist because of the presence of easily ignitable fibers or flyings.

Divisions

The classification applies to each of the three classes and divides them into two divisions according to the degree of hazard based on the conditioning and handling of the substance.

Division 1

This classification identifies the conditions of the area where the hazardous substance is "normally" expected to be present during day-to-day operations. One example of a Class II, Division 1 area is an ammunition process line where explosive gunpowder is used.

Division 2

This classification identifies the conditions of an area where hazardous location substances are *not* normally present in the atmosphere, except under abnormal conditions. One example of a Class I, Division 2 area is when flammable liquids or gases are confined in closed containers or systems where they cannot escape except through an accident.

Groups

Class I, Groups A, B, C, and D are materials divided by flame propagation characteristics (the ability of the fire to multiply), explosion pressures (the force of an explosion if one should occur), and other factors related to the substance that is present or may be present. For example, Group A or Group IIc is an atmosphere containing acetylene.

For the hazardous group atmosphere identifications, see *NEC*® Section 500-5(a) and (b).

Class II, Groups E, F, and G define the ignition temperature (the temperature that will cause spontaneous combustion or self-ignition) and conductivity of each of the types of combustible dusts. For example, Group E is an atmosphere containing aluminum, magnesium, their alloys and other similar combustible dusts. Class III materials are not classified in groups.

Evaluation of Hazardous Areas

Each area containing gases or dusts that are considered hazardous must be carefully evaluated to make certain that the correct electrical equipment is selected. Many hazardous atmospheres are Class I, Group D or Class II, Group G. However, certain areas may involve other groups, particularly Class I, Groups B and C. Conformity with the *NEC®* requires the use of enclosures and fittings approved for the specific hazardous gas or dust involved. In some instances, Group D equipment can be used where Group B atmospheres can be sealed off (see Section 500-5(a)(2), Exception 1).

Classification of Hazardous Areas

Classification of specific areas is determined by the AHJ "code enforcing authority," who may be a representative of the insurance underwriters, a municipal electrical inspector, a fire marshal, or a member of the corporate safety organization.

Even though the professionals who classify locations "live by the Code," many factors may not be obvious. The distance from the hazardous source can affect the classification of the area. Ventilation and air currents must be considered. Temperatures, topography, walls, and special safeguards can also have an effect.

Incorrect classification could lead to using equipment that may not be safe. This can result in fires, explosions, and even loss of life and property. With so many code requirements and location conditions to evaluate, experience, technical knowledge, and judgment are essential for classifying hazardous areas.

Hazardous locations are those areas where a potential for explosion and fire exists because of flammable gases, vapors, or finely pulverized dusts in the atmosphere because of the presence of easily ignitable fibers or flyings. Hazardous locations can result from the normal processing of certain volatile chemicals, gases, grains, and so on or from accidental failure of storage systems for these materials. It is also possible that a hazardous location can be created when volatile solvents or fluids, used in a normal maintenance routine, vaporize to form an explosive atmosphere.

Regardless of the cause of a hazardous location, it is necessary that every precaution be taken to guard against ignition of the atmosphere. Certainly no open flames would be permitted in these locations, but what about other sources of ignition?

For example, I once inspected an ammunition production line where the likelihood of an incident was so high that the designer had the production line celled off with 24-inch concrete walls every 10 to 12 feet. This design would limit the loss and damage to a small area if an explosion occurred. He had also provided many other innovative ideas to minimize dangers in this extremely high-hazard location.

Electrical Sources of Ignition

A source of ignition is simply the energy required to touch off an explosion in a hazardous location atmosphere (see Figure 3–4). When these three elements—oxygen, fuel, and ignition—are present, a fire or explosion will occur. The quantity and volatility of the fuel and oxygen present (fuel-oxygen mix) will determine the extent of the hazard present. The degree of heat required to ignite that mixture depends on the type of fuel present.

For example, the flash point for gasoline is –36°F to –50°F, as compared to the flash point for kerosene, which is 110°F to 162°F. This comparison shows that it might take an open flame to ignite kerosene and only a spark to ignite gasoline. One can see that, generally, fuels requiring an open flame to ignite may be wired in accordance with the first four chapters of the code and not be considered hazardous (fuels with a flash point above 100°F).

Where more hazardous substances are present, it is necessary to consider Chapter 5 methods. However, only when all the factors are known to the designers can they make that final determination for requiring Chapter 5 wiring methods and equipment suitable for hazardous (classified) locations that will minimize the dangers in the day-to-day operations.

The designer must recognize that the electrical equipment can also be a source of this ignition energy. The normal operation of switches, circuit breakers, motor starters, contactors, plugs, and receptacles releases this energy in the form of arcs and sparks as contacts open and close, making and breaking circuits. Electrical equipment such as lighting fixtures and motors are classified as "heat producing," and they will become a source of ignition if they reach a surface temperature that exceeds the ignition temperature of the particular gas, vapor, or dust in the atmosphere.

It is also possible that an abnormality or failure in an electrical system will provide a source of ignition. A loose termination in a splice box or a loose lamp in a socket can be the ignition source of both. The failure of insulation from cuts, nicks, or aging can also act as an ignition source from sparking, arcing, and heat.

Look for the Marking or Label

The *NEC®* contains rules for selection and installation of electrical equipment. Section 500-5(c) and (d) requires equipment for hazardous (classified) locations to be clearly approved and marked with the class, group, and temperature for which it is designed (see Figures 3–5 and 3–6). The Code states that the temperature will be marked as a maximum safe operating temperature or as an identification number, which is listed in Table 500-5(d) of the Code. There are five exceptions to the general rule.

In most cases, there will be a trademark of the testing organization on the label, such as Underwriters Laboratory (UL), Factory Mutual (FM), or Canadian Standards Association (CSA) (see Figure 3–7). These organizations are certifying agencies. Their trademark means that the equipment has been properly tested and has met the standards for the application specified. However, it should be noted that the NFPA committee responsible for Chapter 5 CMP-14 has specified approved equipment rather than listed equipment in many cases. They have substantiated this by stating that much equipment designed and manufactured for hazardous (classified) locations is not mass produced, and therefore listing would substantially increase costs and delay delivery. (Review definitions of "approved" and "listed" in Article 100; many inspectors will require the equipment to be listed before they will approve it.)

Hazardous Locations and the *NEC®*

The *NEC®* treats installations in hazardous locations in Articles 500 through 517. Each hazardous location can be classified by the definitions in the *NEC®*. Following are interpretations of these classifications and applications.

Explosion-Proof Apparatus

An explosion-proof apparatus is equipment that is contained in an approved enclosure marked and labeled for the application. The apparatus is capable of withstanding (containing) an explosion of a specified gas or vapor that might occur within it and of preventing the ignition of a specified gas or vapor surrounding the enclosure by sparks, flashes, or explosion of the gas or vapor that originates within the enclosure. Such an apparatus and enclosure operates so that the external temperature of the enclosure will not reach temperatures that will ignite the surrounding flammable atmosphere (see Figures 3–8 and 3–9).

(d) Marketing: Approved equipment shall be marked to show the class, group, and operating temperature or temperature range referenced to a 40°C ambient.

Exception No. 1: Equipment of the non-heat-producing type, such as junction boxes, conduit, and fittings, and equipment of the heat-producing type having a maximum temperature not more than 100°C (212°F) shall not be required to have a marked operating temperature or temperature range.

Exception No. 2: Fixed lighting fixtures marked for use in Class I, Division 2 or Class II, Division 2 locations only shall not be required to be marked to indicate the group.

Exception No. 3: Fixed general-purpose equipment in Class I locations, other than fixed lighting fixtures, that is acceptable for use in Class I, Division 2 locations shall not be required to be marked with the class, group, division, or operating temperature.

Exception No. 4: Fixed dusttight equipment other than fixed lighting fixtures that are acceptable for use in Class II, Division 2 and Class III locations shall not be required to be marked with the class, group, division, or operating temperature.

Exception No. 5: Electric equipment suitable for ambient temperatures exceeding 40°C (104°F) shall be marked with both the maximum ambient temperature and the operating temperature or temperature range at that ambient temperature.

FPN: Equipment not marked to indicate a division, or marked "Division 1" or "Div. 1," is suitable for both Division 1 and 2 locations. Equipment marked "Division 2" or "Div. 2" is suitable for Division 2 locations only.

The temperature range, if provided, shall be indicated in identification numbers, as shown in Table 500-5(d). Identification numbers marked on equipment nameplates shall be in accordance with Table 500-5(d).

Equipment that is approved for Class I and Class II shall be marked with the maximum safe operating temperature, as determined by simultaneous exposure to the combinations of Class I and Class II conditions.

Table 500-5(d)   Identification Numbers

Maximum Temperature		Identification Number
Degrees Celsius	Degrees Fahrenheit	
450	842	T1
300	572	T2
280	536	T2A
260	500	T2B
230	446	T2C
215	419	T2D
200	392	T3
180	356	T3A
165	329	T3B
160	320	T3C
135	275	T4
120	248	T4A
100	212	T5
85	185	T6

Figure 3–5   Section 500-5(d) and Table 500-5(d). (Reprinted with permission from NFPA 70-1999. *National Electrical Code*®, Copyright © 1998, National Fire Protection Association, Quincy, MA 02269.)

(f) **Class II Temperature:**   The temperature marking specified in (d) shall be less than the ignition temperature of the specific dust to be encountered. For organic dusts that may dehydrate or carbonize, the temperature marking shall not exceed the lower of either the ignition temperature or 165°C (329°F).

FPN: See *Recommended Practice for the Classification of Combustible Dusts and of Hazardous (Classified) Locations for Electrical Installations in Chemical Process Areas,* NFPA 499-1997, for minimum ignition temperatures of specific dusts. (ROP 14-76)

The ignition temperature for which equipment was approved prior to this requirement shall be assumed to be as shown in Table 500-5(f).

Table 500-5(f)

| | Equipment that Is Not Subject to Overloading | | Equipment (Such as Motors or Power Transformers) that May Be Overloaded | | | |
| | | | Normal Operation | | Abnormal Operation | |
Class II Group	Degrees Celsius	Degrees Fahrenheit	Degrees Celsius	Degrees Fahrenheit	Degrees Celsius	Degrees Fahrenheit
E	200	392	200	392	200	392
F	200	392	150	302	200	392
G	165	329	120	248	165	329

Figure 3–6   Section 500-5(f) and Table 500-5(f) (Reprinted with permission from NFPA 70-1999. *National Electrical Code®*, Copyright © 1998, National Fire Protection Association, Quincy, MA 02269.)

Figure 3–7   Examples of labels as they appear on equipment suitable for hazardous locations. (Courtesy of Underwriters Laboratories.)

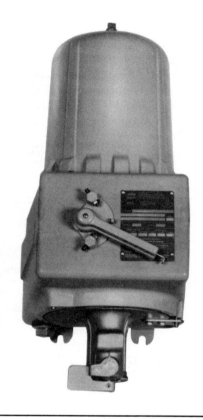

Figure 3–8 Class I switch rack. This
shows the method acceptable for this
hazardous location. (Courtesy of Crouse-
Hinds, Division of Cooper Industries.)

Figure 3–9 Explosion-proof
combination switch and receptacle.
(Courtesy of Crouse-Hinds, Division of
Cooper Industries.)

**Dust-
Ignition-Proof**

NEC® 502-1 states that "as used in this Article" shall mean enclosed in a manner that will
exclude ignitable amounts of dust that might affect performance or rating and that, when
installed and protected in accordance with this code, will not permit arcs, sparks, or heat
otherwise generated or liberated inside the enclosure to cause ignition of exterior accumulations or atmospheric suspensions of a specified dust on or in the vicinity of the enclosure.
Class II equipment is designed to prevent the entrance of dust particles. However, the accumulation of ignitable dust on the enclosure itself is not controllable; therefore, the term
"dust-ignition-proof" means that the enclosure will not reach temperatures high enough
for the dust that accumulates on these enclosures to ignite.

**Class I Zone
Protection
Techniques**

The following protection techniques for electrical and electronic equipment in hazardous
(classified) locations are acceptable in accordance with *NEC*® Article 505 (see Figure
3–10). The equipment, construction, and installation will ensure safe performance under
conditions of proper use and maintenance. For equipment provided with threaded entries
for NPT threaded conduit or fittings, listed conduit, conduit fittings, or cable fittings shall

Example: Class I   Zone 0   AEx   ia   IIC   T6

Area classification ⎯⎯⎯⎯⎯⎯⎯⎯⎯⎯⎯⎯⎯⎯⎯⎯⎯⎯⎯⎯⎯⎯⎯⎯⎯⎯

Symbol for equipment built to American standards ⎯⎯⎯⎯⎯

Type(s) of protection designation ⎯⎯⎯⎯⎯⎯⎯⎯⎯⎯⎯⎯⎯⎯⎯

Gas classification group (not required for protection ⎯⎯⎯⎯
techniques indicated in 505-5, FPN No. 2)

Temperature classification ⎯⎯⎯⎯⎯⎯⎯⎯⎯⎯⎯⎯⎯⎯⎯⎯⎯⎯⎯⎯

Figure 505-10(b)(1)

Table 505-10(b)(1)   Types of Protection Designation

Designation	Technique	Zone 1
d	Flameproof enclosure	1
e	Increased safety	1
ia	Intrinsic safety	0
ib	Intrinsic safety	1
[ia]	Intrinsically safe associated apparatus	Nonhazardous
[ib]	Intrinsically safe associated apparatus	Nonhazardous
m	Encapsulation	1
nA	Nonsparking equipment	2
nC	Sparking equipment in which the contacts are suitably protected other than by restricted breathing enclosure	2
nR	Restricted breathing enclosure	2
o	Oil immersion	1
p	Purged and pressurized	1 or 2
q	Powder filled	1

[1]Does not address use where a combination of techniques is used.
(ROC 14-88)

Table 505-10(b)(2)   Gas Classification Groups

Gas Group	Comment
IIC	See Section 505-5(a)
IIB	See Section 505-5(b)
IIA	See Section 505-5(c)

Figure 3–10   Figure 505-10(b)(1), Table 505-10(b)(1), and Table 505-10(b)(2)—
Some Protection Technique Designations and Gas Classification Groups. (Reprinted with
permission from NFPA 70-1999, *National Electrical Code*®, Copyright © 1998, National
Fire Protection Association, Quincy, MA 02269.)

be used. For equipment provided with threaded entries for metric threaded conduit or fittings, listed adapters to permit connection of NPT threaded conduit, listed conduit fittings, or listed cable fittings or listed cable fittings having metric threads shall be used.

Flameproof "d". This protection technique is permitted for equipment in Class I, Zone 1 locations in which it is approved. This is a type of protection of electrical equipment in which the enclosure will withstand an internal explosion of a flammable mixture. Where it has penetrated into the interior, the internal explosion shall not damage the enclosure and shall not cause ignition through joints or structural openings in the enclosure.

Purged and Pressurized "P". This protection technique shall be permitted for equipment in those Class I, Zone 1 or Zone 2 locations where it is approved. In some cases, hazards may be reduced or hazardous (classified) locations limited or eliminated by adequate positive-pressure ventilation from a source of clean air in conjunction with effective safeguards against ventilation failure. For further information, see *Standard for Purged and Pressurized Enclosures for Electrical Equipment,* NFPA 496-1993.

Encapsulation "m". This protection technique shall be permitted for equipment in those Class I, Zone 1 locations for which it is approved. This is a type of protection in which the parts that could ignite an explosive atmosphere by either sparking or heating are enclosed in a compound in such a way that this explosive atmosphere cannot be ignited. For further information, see *Electrical Apparatus for Explosive Gas Atmospheres—Part 18: Encapsulation* "m," IEC 79-18-1992.

Powder Filling "q". This protection technique shall be permitted for equipment in those Class I, Zone 1 locations for which it is approved. This is a type of protection in which the parts capable of igniting an explosive atmosphere are fixed in position and completely surrounded by filling material (glass or quartz powder) to prevent tile ignition of an external explosive atmosphere.

Intrinsic Safety. This protection technique shall be permitted for equipment in those Class I, Zone 0 or Zone 1 locations for which it is approved. For further information, see *Intrinsically Safe Apparatus and Associated Apparatus in Class I, Class II and Class III Hazardous Locations,* ANSI/LIL 913-1997.

Type of Protection "n". This protection technique shall be permitted for equipment in those Class I, Zone 2 locations for which it is approved. Type of protection "n" is further subdivided into "nA," "nC," and "nR." Type "n" protection is a type of protection applied to electrical equipment such that, in normal operation, the electrical equipment is not capable of igniting a surrounding explosive gas atmosphere and a fault capable of causing ignition is not likely to occur. For further information, see *Electrical Apparatus for Use in Class I, Zone 2 Hazardous (Classified) Locations Type of Protection—"n,"* ISA S12.12.01-1996; and *Electrical Apparatus for Explosive Gas Atmospheres, Part 15—Electrical Apparatus with Type of Protection "n,"* IEC 79-15 (1987).

Oil Immersion "o". This protection technique shall be permitted for equipment in those Class I, Zone 1 locations for which it is approved. Oil immersion is a type of protection in which the electrical equipment or parts of the electrical equipment are immersed in a protective liquid in such a way that an explosive atmosphere that may be above the liquid or outside the enclosure cannot be ignited. For further information, see *Electrical Apparatus for Use in Class I, Zone 1 Hazardous (Classified) Locations, Apparatus for Explosive Gas Atmospheres, Part 6—Oil-Immersion "o,"* IEC 79-6 (1995); *Electrical Equipment for Use in Class I, Zone 0, 1, and 2 Hazardous (Classified) Locations,* UL 2279, 1996; and *Electrical Apparatus for Explosive Gas Atmospheres.*

Hazardous Location

This is an area where the possibility of explosion and fire is created by the presence of flammable gases, vapors, dusts, fibers, or flyings (the latter two being small particles of cloth or wood, often natural and synthetic fibers). These are areas that meet the definition of one of the three classes (Class I, II, or III) as follows.

Class I

Class I is composed of those areas in which flammable gases or vapors may be present in the air in sufficient quantities to be explosive or ignitable (see Appendix A–1a and b and Figures 3–11 and 3–12). These are areas or locations in which the materials are used in production or in day-to-day operations or in which the materials are simply stored and not handled at all.

Class II

Class II is composed of those areas made hazardous by the presence of combustible dust (see Appendix A–2 and Figure 3–13). These are areas or locations similar to the Class I locations in which the materials (combustible dusts) are used in production or in day-to-day operations or in which the materials are simply stored and not handled at all.

Class III

Class III is composed of those areas in which there are easily ignitable fibers or flyings present because of the type of material being handled, stored, or processed (see Figure 3–14). Many believe that the classes are degrees of hazards, which is not true. Each class can be equally as hazardous as the other. Class III has a very high rate of incidence. These areas are typically cabinet shops, furniture factories, textile mills, and so on. We often think of these areas as nonhazardous, but they can be very hazardous if they are not kept very clean. All equipment must be cleaned regularly so that accumulation and buildup of particles and flyings are controlled. Many of the materials that are Class III materials will ignite from the surface temperature of equipment that is in the area but is not specifically approved for the application.

Division 1 in the Normal Situation

This hazard would be expected to be present in everyday production operations or during frequent repair and maintenance activity.

Division 2 in the Abnormal Situation

The material is expected to be confined within closed containers or closed systems and will be present only through accidental rupture, breakage, or unusual faulty operation.

Class I, Zone 0, 1, or 2

Class I can optionally be classified as Class I, Zone 0, 1, or 2. The zone system is suitable for use only for Class I flammable liquids, gases, or vapors.

Class I, Zone 0

These locations are those where ignitable gases or vapors are present in ignitable concentrations continuously or are likely to be present for long periods of time normally.

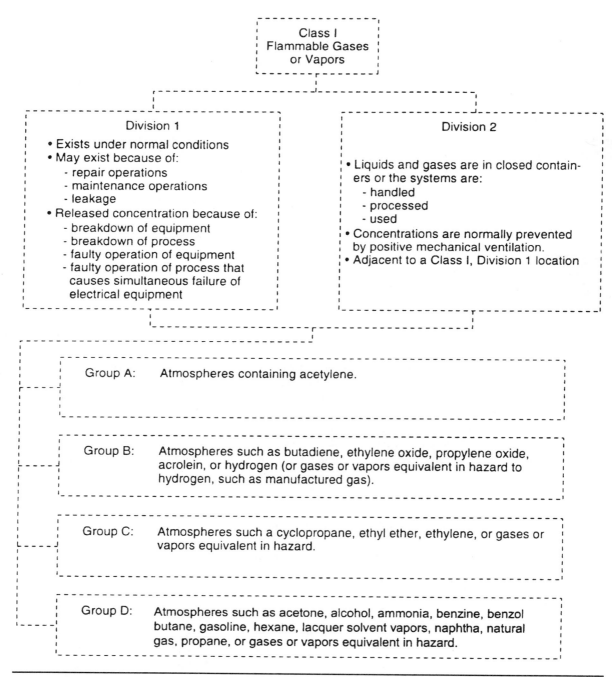

Figure 3–11 Class I classification by class, division, and group in accordance with Article 500.

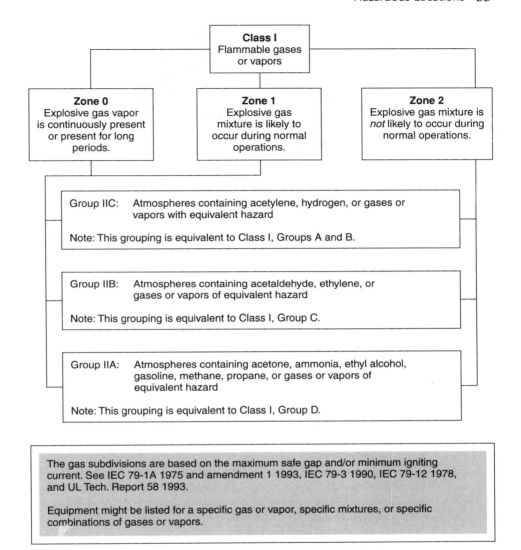

Figure 3–12 Class I classification by class, zone, and group in accordance with NEC Article 505.

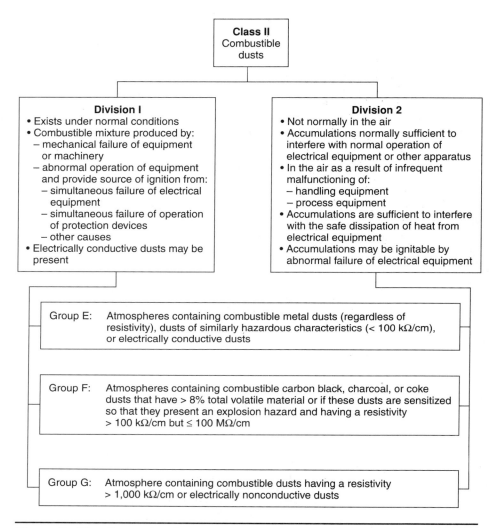

Figure 3–13   Class II classification by class, division, and group in accordance with Article 500.

Class I, Zone 1

These locations are adjacent to Class I, Zone 0 areas where communication of ignitable concentrations of gases or vapors may occur. Where a positive ventilation system from a clean air source with safeguards to prevent ventilation failure is provided, this area may be unclassified. In addition, these are locations where ignitable gases or vapors are present in ignitable concentrations during normal operation or may be present frequently because of repair, maintenance, or leakage.

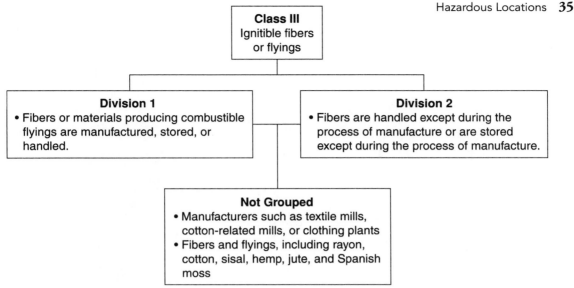

Figure 3–14 Class III classification by class and division in accordance with Article 500.

Class I, Zone 2

These locations are adjacent to Class I, Zone 1 areas where communication of ignitable concentrations of gases or vapors might occur. Where a positive ventilation system from a clean air source with safeguards to prevent ventilation failure is provided, this area may be unclassified. These locations are those where ignitable gases or vapors are not likely to be present in ignitable concentrations during normal operation or might be present for short periods of time. In addition, these are locations where ignitable gases or vapors are present in ignitable concentrations continuously or are likely to be present but are confined in closed containers or processes and can become hazardous only because of rupture, breakdown, or abnormal operation.

Groups

The gases and vapors of Class I locations are broken into groups by the codes A, B, C, or D for the Class I, Division 1 or 2 system and into IIC, IIB, or IIA for the Class I, Zone 0, 1, or 2 system. The hazardous gases and vapors are grouped according to their explosion pressures and other flammable characteristics. Class II locations are grouped as E, F, or G. These groups are classified according to the ignition temperature and the electrical conductivity of the hazardous substance.

Seals

Special fittings are required in Class I locations to minimize the passage of gases and vapors or, in the case of an explosion, to prevent the passage of flames (see Figures 3–15, 3–16, and 3–17). They are also required in Class II locations to prevent the passage of combustible dust.

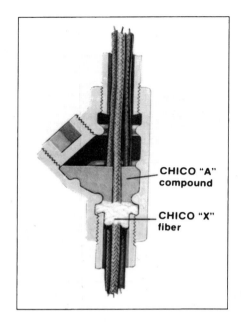

Figure 3–15 Example of an explosion-proof seal fitting suitable for horizontal or vertical installation. (Courtesy of Crouse-Hinds, Division of Cooper Industries.)

Figure 3–16 Cutaway view of an installed vertical seal fitting. (Courtesy of Crouse-Hinds, Division of Cooper Industries.)

Article 500

Article 500 through 503 and 505 (1996 *NEC*®) explains in detail the requirements for the installation of wiring or electrical equipment in hazardous locations. As previously stated here, these articles, along with other applicable regulations, local governing inspection authorities, insurance representatives, and qualified engineering/technical assistance, should be your guides to the installation of wiring or electrical equipment in any hazardous or potentially hazardous location.

EYS – Horizontal seal

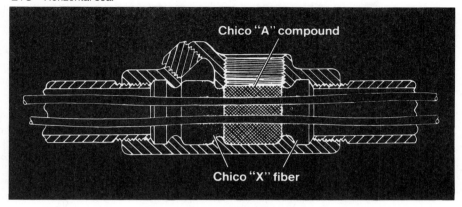

Figure 3–17 Cutaway view of an installed horizontal seal fitting. Two fiber dams are required in this application. (Courtesy of Crouse-Hinds, Division of Cooper Industries.)

Classification of Area

To classify an area requires knowledge of the basic application intended for the installation to be classified. It is important to have a complete understanding of the equipment that is needed and required for the day-to-day operations of that area. It is also important to rely on experience and sound judgment. Electrical area classification is performed to prevent explosions and fires caused by electrical equipment arcing and sparking, high temperatures, or electrical system faults and to prevent personal injury, loss of life, and destruction of buildings and equipment. The installation of correct electrical equipment and materials and the use of the correct installation techniques can make a major contribution toward significantly reducing the risk of such occurrences.

No one individual should be totally responsible for the ultimate classification or the decision for classifying an area. It should be the result of group meetings and agreement with the customer, who can provide experience and the history related to the proposed area or to a similar area using similar hazardous materials. Experienced plant personnel, plant operations, plant maintenance, plant safety department, and in-house engineering should be involved and consulted. Electrical engineers involved with the design, architectural designers involved with the architectural appurtenances, the authority having jurisdiction (which might consist of local and state electrical, fire inspectors, and building officials), and the operating company's underwriters (insurers) should also be consulted. Careful coordination among these groups will not only ensure the correct classification but also provide alternatives to reduce the hazardous areas to a minimum. Many industry codes and standards are available. **Note:** Local codes might need to be consulted as well. These will provide valuable guidance for those ultimately responsible for making these decisions. NFPA-70, the *NEC*®, contains the minimum mandatory electrical installation requirements.

You must first have an industry code that is uniquely related to the industry, chemical, or hazardous material for which this area will be designed, such as API (the American Petroleum Institute). They have unique recommendations for the classification of areas dealing with petroleum facilities, such as API IRP 500-1992. Other industries have similar application codes or guides available. The representative for the operating company should be able to provide this information. There are many standards available through the NFPA® that can help in making this determination.

- NFPA-497 concerns the classification of flammable liquids, gases, or vapors and of hazardous (classified) locations for electrical installations in chemical process areas. This standard has been revised to include former NFPA Standards 497A and a portion of 497M.

- NFPA-499 concerns the classification of combustible dusts and for hazardous (classified) locations for electrical installations in chemical process areas. This standard has been revised to include former NFPA Standards 497B and a portion of 497M.

- NFPA-70, the *NEC®*, Chapter 5, are mandatory requirements as adopted by the jurisdiction in which the facility is being built. You must check with the authority having jurisdiction in that area, state, county, or city to verify which edition of NFPA-70 *NEC®* has been adopted and if there are any amendments.

- NFPA-325-1994 describes the fire hazard for those flammable liquids, gases, and volatile solids. This provides additional information to help determine the gas groups that are not described in NFPA-497-M or elsewhere.

- NFPA-496-1998 deals with purged and pressurized enclosures for electrical equipment.

- ANSI/ISA RP 12.6-1995 deals with intrinsically safe apparatus and associated apparatus for use in Class I, II, and III, Division 1 hazardous locations.

- NFPA-30-1996 is useful for flammable and combustible liquids and covers the ventilation, handling, and storage of such combustible liquids.

- Other standards are available and might be useful in making the proper classification.

After areas have been clearly defined and the class has been established, the group must be established. The groups are listed in alphabetical designations—A through D for Class I, Division 1 or 2 locations. The groups are listed as IIC, IIB, and IIA for Class I, Zone 0, 1, or 2 locations. For Class II, the groups are listed in alphabetical designations E through G. Class III locations are not divided into groups. The NEC, NFPA-70, gives a partial list of the most commonly encountered materials that have been tested for grouping. They are grouped on the basis of their flammability characteristics as Groups A through G, with Group A or IIC for zone classifications being the most volatile (acetylene). Groups E through G are flammable volatile dusts. The responsible party, such as the designer or contractor, should utilize the expertise of chemical process engineers, safety loss prevention engineers, or other experts in developing and establishing the group and class for the area being designed.

Next, the division must be established. *NEC®*-70 recognizes two divisions. A division or zone criterion as based on the presence of volatile liquids, vapors, or dust. A Division 1 location is one that is likely to have volatile liquids, vapors, or dust present under normal conditions, such as an area where the volatile material is dispensed. A Division 2 location is an area that is likely to have flammable or volatile gases, vapors, or dust present only under abnormal conditions, such as accidental spills or accidental releases of the volatile material present.

One should classify the division or zone very carefully. An incorrect area classification of Division 2 or Zone 2 permits a much more liberal wiring method with fewer safety guards built in. Improper classification of an area can greatly reduce the safety factor desired under all operating conditions. It is possible to reduce the classification, but only if there is assurance that all materials are maintained through ventilation and good area janitorial maintenance. However, too often operation management personnel plan on good maintenance and maintaining a clean facility, but within a few months or years the hazards present within the facility (because of poor janitorial maintenance) create a much more hazardous condition than the facility was originally designed for. Existing plant conditions, if available, can provide valuable information to be considered by those responsible during the classification process.

Additional references for the application and use of hazardous materials can be found in Appendix A of NFPA-70 of the *NEC*®. After all available facts are known, a classification is assigned to each building or area. The selection of the materials and the installation must be made in accordance with the adopted edition of *NEC*® and any locally adopted amendments as a minimum installation standard. Many engineers will specify materials and methods more stringent than those in the *NEC*®. All classifying of areas, selection of equipment, and methods of wiring should be under the supervision of a qualified registered professional engineer.

Dual Classification

This might be permitted within the same facility. However, Class I, Zone 2 locations are permitted to abut, but not overlap Class I, Division 2 locations. Class I, Zone 0 or 1 are not permitted to abut Class I, Division 1 or 2 locations.

Reclassification

Class I, Divisions 1 and 2 locations can be reclassified as Class I, Zone 0, 1, or 2 locations provided that there is only a single flammable gas or vapor source and that the area is reclassified in accordance with the provisions of *NEC*® Article 505.

Class I Hazardous (Classified) Areas

Class I Locations

Class I locations as covered in *NEC®* Article 501 or 505 are those in which flammable gases or vapors are or might be present in the air in quantities sufficient to produce explosive or ignitable mixtures (see Figure 4–1).

Class I Location Examples

- Petroleum refining facilities
- Dip tanks containing flammable or combustible liquids
- Dry cleaning plants
- Plants manufacturing organic coatings
- Spray finishing areas (residue must be considered)
- Petroleum dispensing areas
- Solvent extraction plants
- Plants manufacturing or using pyroxylin (nitrocellulose) type and other plastics (Class II also)
- Locations where volatile inhalation anesthetics are used and stored
- Utility gas plants and operations involving storage and handling of liquefied petroleum and natural gas
- Aircraft hangars and fuel servicing areas

Class I Group Classifications

Because the different vapors and gases making up hazardous atmospheres have varying properties, the properties are placed in groups on the basis of common flame propagation characteristics and explosion pressures as defined in *NEC®* Section 500-5(1)-(4). These groups are designated A, B, C, and D for Class I, Divisions 1 and 2 and Groups IIC, IIB, and IIA for Class I, Zones 0, 1 and 2. Reference to the *NEC®* will indicate that much of the equipment used for Class I, Division 2 applications is the same as that used for Division 1

Figure 4–1 Switch rack assembly and lighting in a Class I installation. (Courtesy of Crouse-Hinds, Division of Cooper Industries.)

applications. However, in certain cases, standard location equipment can be used for some of the Class I, Division 2 applications if the appropriate restrictions are followed. The *NEC®* contains the specific rules for these applications. Reference to the *NEC®* Article 505 requires that the equipment used for Class I, Zone 0 be listed and marked suitably for the location.

- Intrinsically safe equipment listed for Class I, Division 1 of the same gas group with suitable temperature rating is also permitted.
- Equipment used for Class I, Zone 1 must be listed and marked suitably for the location; equipment approved for Class I, Division 1 of the same gas group or as permitted by Article 505 and with suitable temperature rating is also permitted.
- Equipment used for Class I, Zone 2 must be listed and marked suitably for the location. Equipment listed for use in Class I, Zone 0 or 1 of the same gas group as permitted by Article 505 and with suitable temperature rating is also permitted. Equipment approved for use in Class I, Division 1 or 2 for the same gas group as permitted by Article 505 and with suitable temperature rating is also permitted. Standard squirrel-cage motors without switching mechanisms, brushes, or other arc-producing devices are permitted. Internal and external surface temperature must be considered.

Class I, Division 1

Class I, Division 1 locations are locations where the hazardous material, flammable gases, or vapors are or might be present in the atmosphere during normal operations. Areas where, during maintenance, repair, breakdown, and faulty operation the presence of ignitable concentrations of gases or vapors can or might be released either continuously, intermittently, or periodically are classified as Division 1 locations. Locations where a breakdown in the operation of processing equipment results in the release of hazardous vapors and the simultaneous failure of electrical equipment are also Class I, Division 1 locations (see Figures 4–2, 4–3, and 4–4).

Class I
Flammable gases and vapors

Condition exists under normal conditions and might
exist during repairs or maintenance.

Division		*Zone*
Group A	Atmosphere containing Acetylene	**Group IIC**
Group B	Atmospheres: Butadiene, ethylene oxide, propylene oxide, acrotein, or hydrogen (also includes gases of equivalent hazard)	**Group IIC**
Group C	Atmospheres: Cyclopropane, ethylether, ethylene, or gases or vapors equivalent in hazard	**Group IIB**
Group D	Atmospheres: Acetone, alcohol, ammonia, benzene, Bengal, butane, gasoline, hexane, lacquer solvent vapors, naphtha, natural gas, propane, or gases or vapors or equivalent hazard	**Group IIA**

Figure 4–2   Example of a Class I location EXGJF or EXLK flexible coupling. Ideal to use with motors and other equipment where flexibility is needed. (Courtesy of Appleton Electric Company.)

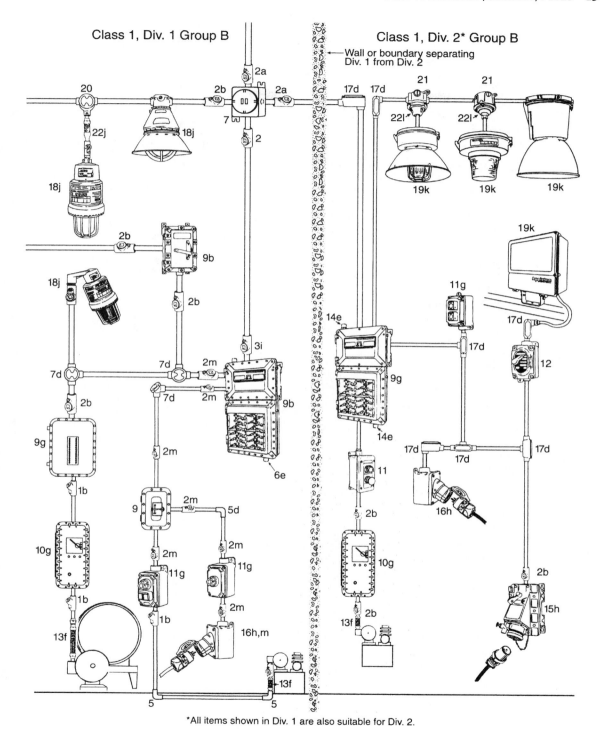

Figure 4–3 Power diagram for a Class I, Division 1 and 2 locations. (Courtesy of Appleton Electric Company.)

Key to Product*

1—Sealing Fitting. EYSF/M, EYS-1,2,3—used with vertical conduits.

2—Sealing Fittings. EYF/M, EYS-11 series, EYS41 series, EYS116 series, EYS416 series—used for sealing vertical or horizontal conduits.

3—Sealing Fitting, expanded fill. EYSEF, EYDEF.

4—Union, UNY-NR, UNF-NR, UNL.

5—Elbows. ELF, ELMF, ELMFL.

6—Drain/Breather. ECDB-B.

7—Explosionproof Junction Boxes. GR, GU, GRH, GRF, ELBY, GRSS, GUBB, with threaded covers. See "m" opposite.

8—Explosionproof Junction Boxes, JBE/JBEW, with ground surface covers.

9—Panelboards, Circuit Breaker, Manual Starter, Disconnect Switch. Div.1—EB, EDS, EWP; Div. 2—D2P.

10—Combination Circuit Breaker and Line Starter. E-Series, bolted cover.

11—Push Button/Pilot Light. Div. 1—EDS, EFDB, OFC; Div.2 only—EFS Contender, UniCode, N2.

12—Switch/Motor Starter, factory sealed. EDS, EFDB.

13—Flexible Coupling. EXGJH, EXLK.

14—Drain/Breather, combination. ECDB-B

15—Receptacle, non-factory sealed, interlocked. FSQX, JBR, EBRH.

16—Receptacle, EFSB, CPC requires seal for Group B, Div. 1. See "m" opposite.

17—Conduit Boxes, Bodies, Fittings. Form 35, 85, 7, 8, Mogul, JB, GSU, LBD.

18—Lighting Fixtures, Div. 1—CodeMaster Jr., AAPA.

19—Lighting Fixtures, Div. 2: HID—Mercmaster III, Mercmaster II, Mercmaster Low Profile, Areamaster I/2, Corroflood.

20—Fixture Hangers, Div. 1—EFHCA.

21—Fixture Hangers, Div. 2—JB, GSU.

22—Flexible Fixture Supports, Div. 1—EXJF; Div. 2—JB Cushion, AHG Cushion, GS Cushion.

*Not every size and style of mentioned series is suitable for Group B. See General Catalog for specific listings.

National Electric Code® Reference

a—*Sec. 501-5(a)(4)*. Seal required (within 10 feet) on either side of boundary entering or leaving hazardous area

b—*Sec. 501-5(a)(1)*. Seals required within 18 inches of all arcing devices.

c—*Sec. 501-5(a)(2)*. Seals required if conduit is 2 inches or larger

d—*Sec. 501-4(a)*. In Div. 1, boxes and fittings must be Group B approved and have 5 full threads engaged with rigid or IMC conduit. In Div. 2, boxes need not be explosionproof. Enclosed and gasketed boxes and wireways are permitted. Approved MI cable and fittings and approved MC cable and fittings are allowed in both Div. 1 and Div. 2.

e—*Sec. 501-5(f)(1)*. Drain/Breathers must be installed to prevent accumulation of liquids or condensed vapors.

f—*Sec. 501-4(a)*. Flexible connections as at motor terminals must be explosionproof and approved for Group B.

g—*Sec. 501-6(a)*. Panelboards, push buttons, switches, motor controllers—shall be explosionproof and approved for Group B.

h—*Sec. 501-12*. Receptacles and plugs must be explosionproof, Group B approved and provide grounding conductor for portable equipment.

i—*Sec.501-6*. Approved expanded fill seals permit up to 40% fill of cross sectional area of conduit.

j—*Sec 501-9*. Lighting fixtures in Div. 1 must be listed and if stem exceeds 12 inches it must be braced or have a flexible connector.

k—*Sec. 501-9(b)*. In Div. 2, fixtures must be tested and marked. Hot spot can not exceed 80% of ignition temperature in degrees C of gas /vapor involved.

l—*Sec 501-9(b)(3)*. In Div.2, hangers must be approved for Div. 2. If stem exceeds 12 inches it must be braced or have a flexible connector.

m—For Div. 1, Group B listings, some products require seals installed <u>immediately adjacent.</u> See catalog for exact distance (can vary by product).

Figure 4–3 *Continued.*

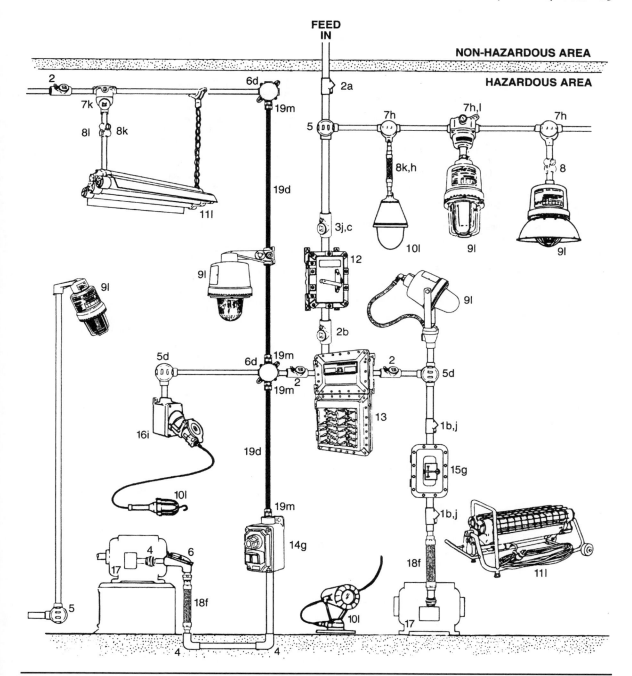

Figure 4–4 Lighting diagram for a Class I, Division 1 location. (Courtesy of Appleton Electric Company.)

Key to Product

1—Sealing Fittings. EYSF/M, EYS 1, 2, 3, 16, 26, 36—used with vertical conduits.

2—Sealing Fittings. EYF/M, EYS, EYD, EYDM, ESUF/M—used with vertical or horizontal conduits.

3—Sealing Fittings, expanded fill. EYSEF, EYDEF.

4—Unions/Elbows. UNY-NR, UNF-NR, UNY/F, UNL, UNYL/UNFL; ELF, ELMF.

5—Explosionproof Junction Boxes. GR, GRSS, GRF, GUBB, GRU, GRUE, GU, with threaded covers.

6—Explosionproof Junction boxes. ELBD, CPU, JBE, JBEW, with ground surface covers.

7—Fixture Hangers. CPU, EFHU, EFHC, GRF.

8—Flexible Fixture Supports. EXJF, ESD, ESS.

9—Lighting Fixtures, HID. CodeMaster-2, Codemaster Jr., CodeMaster Flood.

10—Lighting Fixtures, Incandescent. A-51, EHL, G-EFWB.

11—Lighting Fixtures, Fluorescent. EFU, APL, PAPL.

12—Circuit Breaker. EB.

13—Panelboard, factory sealed. EWP.

14—Push Button/ Pilot Light, factory sealed. EDS.

15—Switch/Manual Motor Starter. EXMS, EDS Disconnect.

16—Receptacle, factory sealed, U-Line. EFS.

17—Motor for Hazardous Location.

18—Flexible Connector. EXGJH, EXLK.

19—MC Cable/Connectors, listed for Class I, Div. 1.

*Also suitable for Class I, Zone 1, see 505-15(b)

National Electric Code® Reference

a—*Sec. 501-5(a)(4)*. Seal required on either side of boundary (within 10 feet) entering or leaving hazardous area.

b—*Sec. 501-5(a)(1)*. Seals required within 18 inches of all arcing devices.

c—*Sec. 501-5(a)(2)*. Seals required if conduit is 2 inches or larger.

d—*Sec. 501-4(a)*. Boxes must be explosionproof and have 5 full threads engaged with rigid or IMC conduit. Approved MI cable and fittings allowed. Approved MC cable allowed with approved cable connectors.

e—*Sec. 501-5(f)(1)*. Drain/breathers must be installed to prevent accumulation of liquids or condensed vapors.

f—*Sec. 501-4(a)*. Flexible connections as at motor terminals must be explosionproof.

g—*Sec. 501-6(a)*. Push buttons, switches, motor controllers, shall be explosionproof.

h—*Sec. 501-9 (a)(4)*. Boxes and fittings used for support of lighting fixtures shall be approved for Class I.

i—*Sec. 501-12*. Receptacles and plugs must be explosionproof and provide grounding conductor for portable equipment.

j—*Sec.501-6*. Approved expanded fill seals permit up to 40% fill of cross sectional area of conduit.

k—*Sec. 501-9(b)(3)*. Pendant fixture stems must be must be threaded rigid or IMC conduit. Stems over 12 inches must be braced or have approved flexible connector.

l—*Sec. 501-9(a)(1)*. All lighting fixtures, both fixed and portable, must be approved for Class I, Div. 1.

m—Appleton TMCX MC Connectors listed for Class I, Div. 1.

Figure 4–4 *Continued.*

Examples of Class I, Division 1 Locations

- Locations where volatile flammable liquids or liquefied flammable gases are transferred from one container to another
- Paint spray booths
- Areas where there are open vats or tanks of flammable liquids
- Drying rooms or compartments for the evaporation of flammable liquids
- Portions of dyeing or cleaning plants where flammable liquids are used
- Gas generation rooms
- Inadequately ventilated pump rooms for flammable liquids
- Interiors of refrigerators and freezers where flammable liquids are stored
- Locations where flammable liquids or gases might be present in ignitable concentrations during normal operations

Class I, Division 2

Class I, Division 2 locations are those in which volatile flammable liquids or gases are handled, processed, or used but in which they will normally be confined within closed con-

Figure 4–5   Example of Class I, Division 2 location lighting in a refinery. (Courtesy of Crouse-Hinds, Division of Cooper Industries.)

tainers or closed systems from which they can escape only in the case of accidental rupture or breakdown of the containers or systems (see Figure 4–5). The hazardous conditions will occur only under abnormal conditions. Class I areas can also be classified as Division 2 where ignitable concentrations of flammable gases or vapors are prevented by positive mechanical ventilation where safeguards against ventilated failure are provided.

Electrical conduits and their associated enclosures separated from process fluids by a single seal or barrier must be classed as a Division 2 location if the outside of the conduit and enclosures is an unclassified location.

Examples of Class I, Division 2 Locations

- A location where volatile flammable liquids, gases, or vapors are used but in the judgment of the authority having jurisdiction would become hazardous in the case of an accident or unusual operating condition. (See Figures 4–6 and 4–7.)

- A location where volatile flammable liquids, gases, or vapors are used but in the judgment of the authority having jurisdiction adequate ventilating equipment with safeguards are used to ensure that the ventilation equipment will provide continuous ventilation air movement.

Factors for Determining Division 2 or Zone 2 Areas

- Limited quantity of flammable material that might escape in an accident or spill
- Adequacy of ventilating equipment
- Specific industry or business records, as they relate to previous explosions and fires

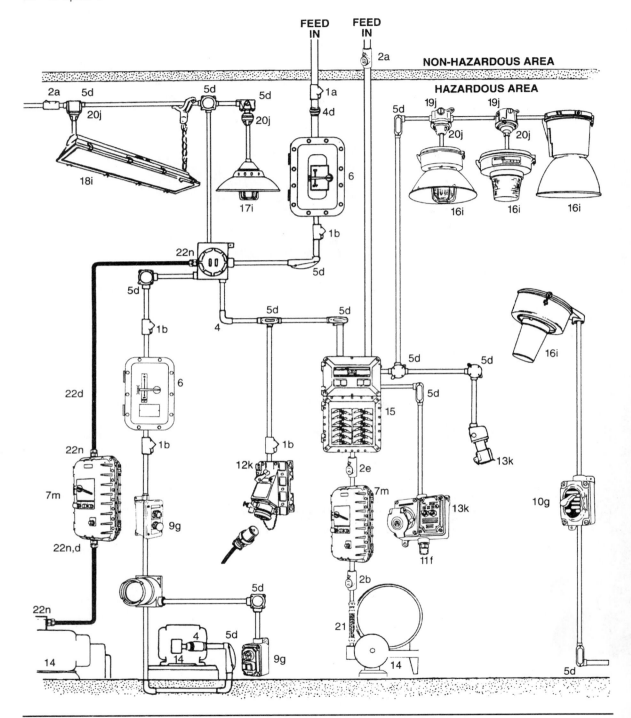

Figure 4–6 Lighting diagram for a Class I, Division 2 location. (Courtesy of Appleton Electric Company.)

Key to Product

1—Sealing Fitting. EYSF/M, EYS 1, 2, 3, 16, 26, 36—used with vertical conduits.

2—Sealing Fittings. EYF/M, EYS, EYD, EYDM, ESUF/M—used with vertical or horizontal conduits.

3—Sealing Fitting, expanded fill. EYSEF, EYDEF.

4—Unions/Elbows. UNY-NR, UNF-NR, UNY/F, UNL, UNYL/UNFL; ELF, ELMF.

5—Conduit Boxes, Bodies, Fittings. Form 35, 85, 7, 8, Mogul, JB, GSU, LBD, RS.

6—Circuit Breaker or Disconnect Switch. EB, E-Series, EXMS, EDS.

7—Combination Circuit Breaker and Line Starter. E-Series, bolted cover.

8—Combination Line Starter and Circuit Breaker, E Series, threaded cover.

9—Push Button/Pilot Light, factory sealed. EDS, EFDB, UniCode.

10—Switch/Motor Starter, factory sealed. EFD, EFDB, EDS.

11—Drain/Breather, combination. ECDB.

12—Receptacle, non-factory sealed, interlocked, EBR, JBR.

13—Receptacle, factory sealed, U-Line. EFS with GFI, CES, CESD.

14—Motor for Div. 2 Hazardous Location.

15—Lighting Panelboard, factory sealed. D2P.

16—Lighting Fixtures, HID—Mercmaster III, Mercmaster II, Mercmaster Low profile, Areamaster I/2, Corroflood.

17—Lighting Fixtures, Incandescent—Stylmaster, V-51.

18—Lighting Fixtures, Fluorescent—VRS, Mercmaster III Compact Fluorescent.

19—Fixture Hangers. JB, GSU.

20—Flexible Fixture Supports. AHG, EXJF, JB Cushion, AHG Cushion

21—Flexible Connector. EXGJH, EXLK; Liquidtight flexible metal conduit and fittings.

22—Cable/Connectors. MC—listed for Div. 2.

* Also suitable for Class I, Zone 2. See 505-15(c)

National Electric Code® Reference

a—*Sec. 501-5(a)(4).* Seal required on either side of boundary entering or leaving hazardous area.

b—*Sec. 501-5(a)(1).* Seals required within 18 inches of all arcing devices.

c—*Sec. 501-5(a)(2).* Seals required if conduit is 2 inches or larger.

d—*Sec. 501-4(b).* Boxes, fittings and joints not required to be explosionproof. Wiring methods shall be threaded rigid metal conduit or steel IMC. Also a variety of cable and raceway systems are permitted. Approved MC cable allowed with approved cable connectors.

e—*Sec 501-5(b)(1).* Seals required in each conduit run entering and leaving an enclosure containing an arcing device.

f—*Sec. 501-5(f)(1).* Drain/breathers must be installed to prevent accumulation of liquids or condensed vapors.

g—*Sec. 501-6(a).* Push buttons, switches, motor controllers, shall be explosionproof.

h—*Sec. 501-9(b)(1).* Portable lighting must be approved for Class I Div.1.

i—*Sec. 501-9 (b)(2).* Fixed lighting shall be enclosed and gasketed and marked to show maximum operating temperature. Hot spot cannot exceed 80% of ignition temperature in degrees C of gas/vapor involved.

j—*Sec. 501-9(b)(3).* Pendant fixtures shall be suspended by rigid metal or IMC conduit. Bracing or flexible connections required if stem exceeds 12 inches.

k—*Sec. 501-12.* Receptacles and plugs must be explosionproof and have grounding conductor for portable equipment.

l—*Sec. 501-5(c)(6).* Approved expanded fill seals permit up to 40% fill of cross sectional area of conduit.

m—*Sec. 501-6(b)(1).* Switches, motor controllers, circuit breakers, and fuses must be in enclosures approved for Class I, Div. 1.

n—Listed MC Cable connector, TMCX.

Figure 4–6 *Continued.*

Equipment for Class I, Division 1 Locations

Equipment used in Class I, Division 1 locations must be marked in accordance with *NEC®* Section 500-5(d) with the following:

- Class
- Group
- Operating temperature or temperature range referenced to 40°C ambient

Devices for Class I locations are housed in enclosures that are designed strong enough to contain an explosion if the hazardous vapors enter the enclosure and are ignited. These enclosures then cool and vent the products of combustion in such a way that the surrounding atmosphere is not ignited. It is important to note that equipment designed and

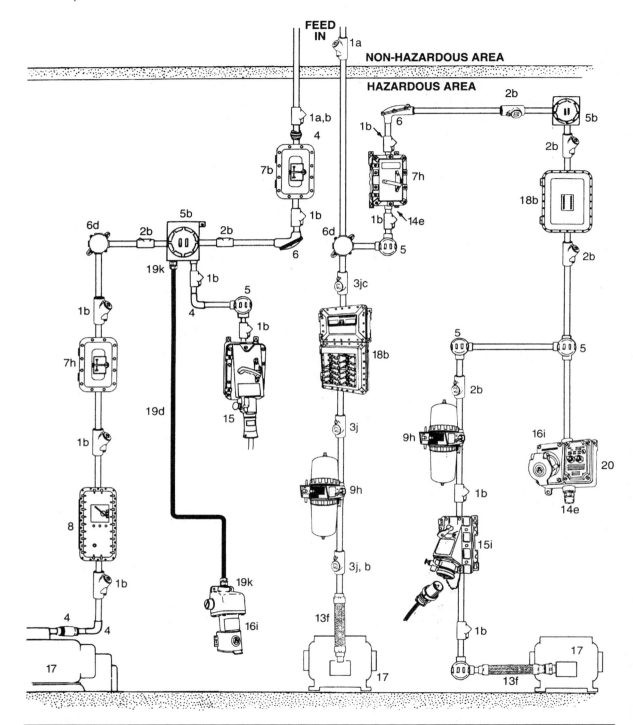

Figure 4–7 Power diagram for a Class I, Division 2 location. (Courtesy of Appleton Electric Company.)

Key to Product

1—Sealing Fitting. EYSF/M, EYS 1, 2, 3, 16, 26, 36—used with vertical conduits.
2—Sealing Fittings. EYF/M, EYS, EYD, EYDM, ESUF/M—used with vertical or horizontal conduits.
3—Sealing Fittings, expanded fill. EYSEF, EYDEF.
4—Unions/Elbows. UNY-NR, UNF-NR, UNY/F UNL, UNYL/UNFL; ELF, ELMF.
5—Explosionproof Junction Boxes. GR, GRSS, GRF, GUBB, GRU, GRUE, GU, ELBY, with threaded covers.
6—Explosionproof Junction Boxes. ELBD, CPU, ER, JBE, JBEW, with ground surface covers.
7—Circuit Breaker/Disconnect Switch/Manual Starters. E-Series, EB, EXMS, EDS, N1.
8—Combination Circuit Breaker and Line Starter. E-series bolted cover.
9—Combination Line Starter and Circuit Breaker. E-series threaded cover.
10—Push Button/Pilot Light, factory sealed. EDS, EFDB, EFD, EFS/D Contender.
11—Push Button, non-factory sealed. N1, EFD, OFC.
12—Switch/Motor Starter, factory sealed. EFD, EDS, EFDB.
13—Flexible Connector. EXGJH, EXLK.
14—Drain/Breather, combination. ECDB.
15—Receptacle, non-factory sealed. FSQX, JBR, EBR.
16—Receptacle, factory sealed. U-Line, EFS, CPS, CES, CESD.#
17—Motor for Explosionproof Location.
18—Panelboards. EWP, EPB, GHB, FB.
19—Cable/Connectors, MC—must be approved for Class I, Div. 1.
20—Ground Fault Interrupter (GFI)

*Also suitable for Class I, Zone 1, see 505-15(b)
60 Amp CES/CESD Suitable for Group D only

National Electric Code® Reference

a—*Sec. 501-5(a)(4).* Seal required on either side of boundary (within 10 feet) entering or leaving hazardous area.
b—*Sec. 501-5(a)(1).* Seals required within 18 inches of all arcing devices.
c—*Sec. 501-5(a)(2).* Seals required if conduit is 2 inches or larger
d—*Sec. 501-4(a).* Boxes must be explosionproof and have 5 full threads engaged with rigid or IMC conduit. Approved MI cable and fittings allowed. Approved MC cable allowed with approved cable connectors.
e—*Sec. 501-5(f)(1).* Drain/breathers must be installed to prevent accumulation of liquids or condensed vapors.
f—*Sec. 501-4(a).* Flexible connections as at motor terminals must be explosionproof.
g—*Sec. 501-4(a).* All boxes, fittings and joints shall be explosionproof.
h—*Sec. 501-6(a).* Circuit breakers, push buttons, switches, motor controllers—shall be explosionproof.
i—*Sec. 501-12.* Receptacles and plugs must be explosionproof and provide grounding conductor for portable equipment.
j—*Sec. 501-6.* Approved expanded fill seals permit up to 40% fill of cross sectional area of conduit.
k—Listed MC Cable Connector, TMCX

Figure 4–7 *Continued.*

listed or approved to be installed in these locations is *not* gas-tight. It *is* designed to contain the explosive forces and internal pressures if an explosion should occur.

Heat-producing equipment for hazardous locations, such as lighting fixtures, must not only contain the explosion and vent the cooled products of combustion but also be designed to comply with *NEC®* Section 500-5(e) and operate with surface temperatures below the ignition temperatures of the hazardous atmosphere (see Figures 4–4 and 4–8).

Specific Equipment for Division 1 Locations

Transformers and capacitors that do not contain liquids that will burn must be installed in vaults or be approved for Class I installations. Transformers and capacitors that do contain liquids that burn must be installed in vaults complying with *NEC®* Sections 450-41 through 450-48 with adequate ventilation for the continuous removal of flammable gases and vapors. Sufficient venting is required to relieve explosion pressures if an explosion should occur. Doors or openings are not permitted between the Class I area and the vault (see Figure 4–3).

Design of Explosion-Proof Equipment

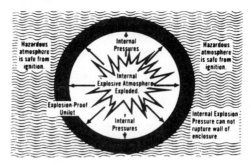

Threaded Joint Construction

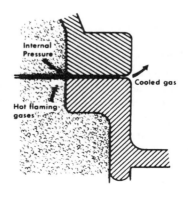

Figure 4–8 Explosion-proof construction is not gas-tight by design. It is designed and manufactured to contain an explosion and to prevent the escape of fire and hot gases. (Courtesy of Appleton Electric Company.)

Meters, instruments, and relays, such as kilowatt-hour meters, thermionic tubes, and instrument transformers, must be provided with enclosures approved for Class I, Division 1 locations.

Switches, motor controllers, circuit breakers, fuses, push-buttons, relays, and similar devices must be provided with enclosures, and together they must be approved as a complete assembly for the Class I, Division 1 location. Control transformers, impedance coils, and resistors, including associated switching mechanisms and their enclosures, must be approved as a complete assembly for the Class I, Division 1 location (see Figure 4–3).

Motors, generators, and other rotating electric machinery must be approved for Class I, Division 1 locations or meet the special manufacturing and installation requirements as specified in *NEC®* Section 501-8(a).

Lighting fixtures must be approved as a complete assembly for Class I, Division 1 locations and clearly marked for the maximum lamp wattage permitted. All fixtures are required to have physical protection either by location or by suitable guards. All boxes, box assemblies, or fittings used to support fixtures must be approved for Class I, Division 1 locations. Portable fixtures must be approved for portable use and comply with the above requirements (see Figures 4–9 and 4–10).

Equipment for Class I, Division 2 Locations

Equipment used in Class I, Division 2 locations must be marked in accordance with the *NEC®* with the following:

- Class
- Group
- Operating temperature or temperature range referenced to 40°C ambient

Fixed lighting fixtures are not required to be marked to indicate the group. Fixed general purpose equipment, except fixed lighting fixtures, acceptable for Class I, Division 2, are not required to be marked with class, group, division, or operating temperature.

Devices for Class I, Division 2 locations are housed in enclosures that are designed strong enough to contain an explosion if the hazardous vapors enter the enclosure and are ignited. These enclosures then cool and vent the products of combustion in such a way that the surrounding atmosphere is not ignited. It is important to note that equipment designed and listed or approved to be installed in these locations is *not* gas-tight. It *is* designed to contain the explosive forces and internal pressures if an explosion should occur (see Figure 4–7). However, if all the provisions of Section 501-6(b)(1) through (4) are met, using general purpose enclosures to house certain equipment may be permitted.

Specific Equipment for Division 2 Locations

Transformers and capacitors must comply with *NEC®* Article 450. Meters, instruments, and relays, such as kilowatt-hour meters, thermionic tubes, and instrument transformers, must be provided with enclosures approved for Class I, Division 1 locations.

Switches, circuit breakers, fuses, make and break contacts of push-buttons, relays, and similar devices located in a Class I, Division 2 area must meet the requirements of the *NEC®* and be provided with enclosures, and together they must be approved as a complete assembly for the Class I, Division 1 location. Control transformers, impedance coils, and resistors, together with switching mechanisms and their enclosures, must be approved as a complete assembly for the Class I, Division 1 location. General purpose enclosures are permitted, provided that the current-interrupting contacts are immersed in oil, enclosed

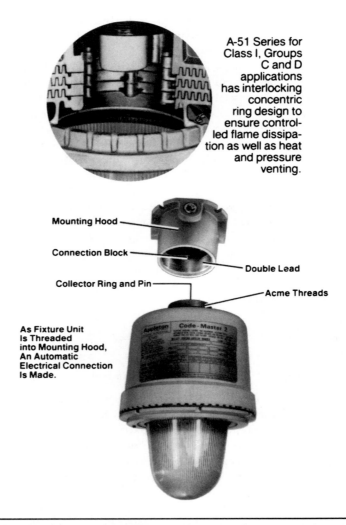

A-51 Series for Class I, Groups C and D applications has interlocking concentric ring design to ensure controlled flame dissipation as well as heat and pressure venting.

Mounting Hood

Connection Block

Double Lead

Collector Ring and Pin

Acme Threads

As Fixture Unit Is Threaded into Mounting Hood, An Automatic Electrical Connection Is Made.

Figure 4–9 Explosion-proof light fixtures. (Courtesy of Crouse-Hinds, Division of Cooper Industries.)

within a chamber that is hermetically sealed to prevent the entrance of gases or vapors, or in nonincendive circuits which under normal conditions do not release enough energy to ignite the specific ignitable atmospheric mixture (see Figure 4–7).

Motor controllers, motors, generators, and other rotating electric machinery located in Class I, Division 2 areas must be approved for Class I, Division 1 locations or meet the special manufacturing and installation requirements as specified in *NEC®* Section 501-8(b).

Equipment used in Class I, Division 2 locations must be marked in accordance with the *NEC®* with the following:

- Class
- Group
- Operating temperature or temperature range referenced to 40°C ambient

Figure 4–10   Portable reel-lights suitable for use in Class I, Divisions 1 and 2. (Courtesy of Appleton Electric Company.)

Wiring Methods

The minimum wiring methods are specified in *NEC®* Article 501, Section 501-4(a) for Class I, Division 1 locations. *NEC®* Section 501-4(b) covers Class I, Division 2 locations. Class I, Zones 0, 1, and 2 wiring methods are covered in Section 505-15.

Class I, Division 1

NEC® 501-4(a) requires wiring methods of *NEC®* Article 346 Threaded Rigid Metal Conduit, *NEC®* Article 345 Threaded Steel Intermediate Metal Conduit (IMC), or *NEC®*

Article 330 type MI cable with termination fittings approved for the location. MI cable must be installed and supported in a manner to avoid tensile stress at the terminations. All boxes, fittings, and joints must be threaded for a connection to conduit or cable terminations and must be explosion-proof. Threaded joints must be made up with at least five threads fully engaged (see Figures 4–9, 4–12, and 4–14).

Where necessary to use flexible connections as provided in *NEC®* Section 501-11, such as for motor terminals, flexible fittings approved for Class I locations must be used (see Figure 4–10). *NEC®* Section 501-11, however, permits flexible cord or connection between portable lighting equipment or other portable utilization equipment and a fixed portion of its supply circuit with rigid requirements. Extra-hard usage cord must be employed and must, in addition to the conductors in the circuit, employ a grounding conductor complying with *NEC®* Section 400-23. It must be connected to terminals or supply conductors in an approved manner and be supported by clamps or other suitable means in such manner that there will be no tension on the terminal connections (see Figure 4–13). The cord and connections must be provided with seals where the cord enters boxes, fittings, or enclosures of explosion-proof type.

In industrial establishments with restricted public access and where conditions of maintenance and supervision assure that only qualified persons will service the installation, type MC (*NEC®* Article 334) and type ITC (*NEC®* Article 727) can be used, provided that the cable type is listed for the use in Class I, Division 1 locations. The cable must have a gas- and vapor-tight continuous corrugated aluminum sheath and an overall jacket of suitable polymeric material. The termination fittings must be listed for the use. The type MC cable must also contain an equipment-grounding conductor sized in accordance with *NEC®* Section 250-122 (see Figure 4–18).

Rigid nonmetallic conduit (Article 347) is permitted where encased in a minimum of 2 inches of concrete and buried below the surface at least 2 feet. Threaded rigid metal conduit or intermediate metal conduit shall be used for the last 2 feet of the run and to the point of emergence or point of connection to above-ground raceways. An equipment-grounding conductor shall be provided in accordance with Section 250-122 to provide continuity of the raceway system.

Class I, Zone 0

Only the intrinsically safe wiring in accordance with *NEC®* Article 504 methods is permitted. Seals must be provided within 10 feet of the Zone 0 boundary where the conduit leaves the area. The seals are not required to be explosion-proof or flameproof. Seals are required at the first point of termination inside the Zone 0 area. Where rigid conduit passes through the Zone 0 area with no fittings closer than 12 inches of a boundary and the termination points are in a unclassified area, the conduit is not required to be sealed.

Class I, Zone 1

All wiring methods permitted for Class I, Division 1 locations are acceptable in Zone 1 locations. The wiring methods must be sealed and drained in accordance with *NEC®* Article 501. An explosion-proof seal must be provided for each conduit entering enclosures having protection types e or d. Protection, except for type d enclosures that are marked to indicate a seal, is not required. Wiring methods must maintain the integrity of protection techniques (see figure 4–11).

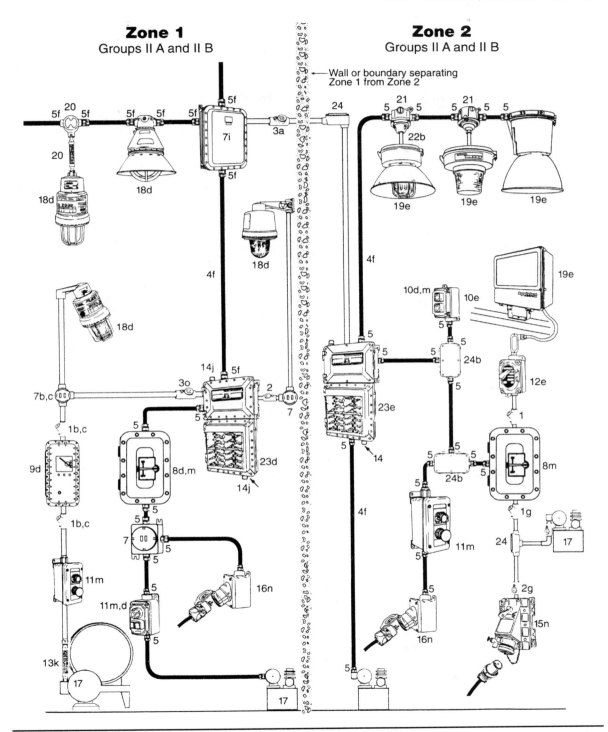

Figure 4–11   Class I, Zones 1 and 2 lighting and power diagram. (Courtesy of Appleton Electric Company.)

Key to Product

1—Sealing Fittings. EYSF/M, EYS 1, 2, 3, 16, 26, 36—used with vertical conduits.

2—Sealing Fittings. EYF/M, EYS, EYD, EYDM, ESUF/M—used with vertical or horizontal conduits.

3—Sealing Fittings, expanded fill. EYSEF, EYDEF.

4—Cable, MC—must be approved for Zone 1 and Zone 2.

5—Cable connector, TMCX. See "f" opposite.

6—Unions. UNY-NR, UNF-NR, UNY/F UNL, UNYL/UNFL

7—Explosionproof Junction Boxes. GR, GRSS, GRF, GUBB, GRU, GRUE, JBE/JBEW.

8—Circuit Breaker or Disconnect Switch. E Series, EB, EXMS, EDS.

9—Combination Circuit Breaker and Line Starter. E-Series, with bolted cover.

10—Push Button/Pilot Light, factory sealed. EDS, EFDB.

11—Push Button/Pilot Light, factory sealed, approved for Zone 1 and Zone 2. UniCode, EDS, EFD.

12—Switch/Motor Starter, factory sealed. EFD, EFDB.

13—Flexible Coupling. EXGJH, EXLK.

14—Drain/Breather, combination. ECDB.

15—Receptacle, non-factory sealed, interlocked. EBR, JBR.

16—Receptacle, factory sealed. U-Line, EFS.

17—Motor for explosionproof location.

18—Lighting Fixtures, Zone 1. CodeMaster, CodeMaster Jr., A-51, EFU.

19—Lighting Fixtures, Zone 2. Mercmaster III, Mercmaster II, Mercmaster Low profile, Areamaster I/2, Corroflood.

20—Fixture Hangers, Zone 1. EXJF, EFHC, EFHU.

21—Fixture Hangers, Zone 2. JB, GSU.

22—Flexible Fixture Supports, Zone 2. JB Cushion, AHG Cushion.

23—Panelboard, Zone 1. EWP; Zone 2–D2P.

24—Conduit Boxes, Bodies, Fittings, Zone 2. Form 35, 85, 7, 8, Mogul, JB, GSU, LBD, RS.

National Electric Code® Reference

a—*Sec. 501-5(a)(4)*. Seal required on either side of boundary entering or leaving hazardous area.

b—*Section 505-15*. In Zone 1, all wiring methods required for Class 1, Div. 1 are also required here. Similarly, in Zone 2, all wiring methods required for Div. 2 are also required.

c—*505-15(2)*. Listed explosionproof fittings and boxes used with rigid or steel IMC required.

d—*Section 505-20(b)*. In Zone 1, equipment must be specifically listed and marked. Equipment listed for Class 1. Div.1 of the same gas group and with similar temperature marking (if any) is permitted.

e—*Section 505-20(c)*. In Zone 2, equipment suitable for Class 1, Div. 1 or Div. 2 of the same gas group and similar temperature is permitted.

f—*Section 505-15(3)*. Approved MC cable and listed fittings are suitable for use in Zone 1 and Zone 2.

g—*Sec. 501-5(a)(1)*. Seals required within 18 inches of all arcing devices.

h—*Sec. 501-5(a)(2)*. Seals required if conduit is 2 inches or larger.

i—*Sec. 501-4(a)*. Boxes must be explosionproof and have 5 full threads threads engaged with rigid or IMC conduit. Approved MI cable and fittings allowed. Approved MC cable allowed with approved cable connectors.

j—*Sec. 501-5(f)(1)*. Drain/breathers must be installed to prevent accumulation of liquids or condensed vapors.

k—*Sec. 501-4(a)*. Flexible connections as at motor terminals must be explosionproof.

l—*Sec. 501-4(a)*. All boxes, fittings and joints shall be explosionproof.

m—*Sec. 501-6(a)*. Push buttons, switches, motor controllers. shall be explosionproof.

n—*Sec. 501-12*. Receptacles and plugs must be explosionproof and provide grounding conductor for portable equipment.

o—*Sec.501-5(c)(6)*. Approved expanded fill seals permit up to 40% fill of cross sectional area of conduit.

Figure 4–11 *Continued.*

Figure 4–12 Explosion-proof flexible conduit is approved for limited use where necessary for flexibility. Use only approved wiring methods pictured above. (Courtesy of Crouse-Hinds, Division of Cooper Industries.)

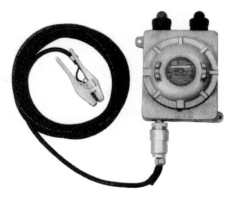

Figure 4–13 Electrical equipment with a static ground clamp. (Courtesy of Crouse-Hinds, Division of Cooper Industries.)

Figure 4–14 The example of a thread lubricant manufactured for the purpose; only approved thread lubricants should be used. (Courtesy of Crouse-Hinds, Division of Cooper Industries.)

Class I, Zone 2

All wiring methods permitted for Class I, Divisions 1 and 2 locations are acceptable in Zone 2 locations. The wiring methods must be sealed and drained in accordance with *NEC®* Article 501. Wiring methods must maintain the integrity of protection techniques (see figure 4–11).

Class I, Division 2

NEC® Section 501-4(b) requires somewhat less stringent wiring requirements. It is equally important, however, to wire areas classified as Class I, Division 2 correctly. The Class I, Division 2 minimum wiring methods are as follows: *NEC®* Article 346 Threaded Rigid Metal Conduit, *NEC®* Article 345 Threaded Steel Intermediate Metal Conduit, *NEC®* Article 364 Enclosed Gasketed Busways, *NEC®* Article 362 Enclosed Gasketed Wireways, PLTC cable in accordance with the provisions of *NEC®* Article 725, and types MI, MC, MV, or PC cable with approved fittings and terminations are permitted. Types PLTC, MI, MC, MV, TC, and SNM are permitted to be installed in cable trays and must be installed in a manner to avoid tensile stress at the termination fittings.

Boxes and joints are not required to be explosion-proof in these areas, except as required in *NEC®* Sections 501-3(b)(1), 501-6(b)(1), and 501-14(b)(1). When provisions must be made for limited flexibility, such as at motor terminals, flexible metal conduit with approved fittings, *NEC®* Article 333 type AC Armored Cable with approved fittings, *NEC®* Article 351 Liquidtight Flexible Metal Conduit with approved fittings, and Liquidtight Flexible Nonmetallic Conduit with approved fittings, or flexible cord approved for hard uses provided with approved bushed fittings can be used (see Figure 4–13). The additional conductor for grounding must be included in the flexible cord unless other accessible means for grounding are provided.

Wiring that cannot release sufficient energy to ignite a specific ignitable atmospheric mixture by opening, shorting, or grounding shall be permitted in this area, using any of the methods suitable for general wiring applications. Nonincendive circuits are permitted in wiring methods suitable for ordinary locations.

Sealing and Drainage Requirements

NEC® Section 501-5 covers the minimum requirements for sealing and drainage in Class I, Divisions 1 and 2. Fittings approved for the location and purpose must be used with approved sealing compounds; sealing compounds approved for type MI cable must also exclude moisture and other fluids from the cable insulation. In addition, see the requirements of *NEC®* Section 505-15 for zone system sealing requirements.

Seals are provided in conduit and cable systems to minimize the passage of gases and vapors and prevent the passage of flames through the wiring method from one electrical system or enclosure to another. Type MI cable inherently prevents the passage of gases and vapors through the insulation; therefore, seals are not required.

The specific requirements for installing the seals are found in *NEC®* Section 501-5(c) and in the manufacturer's instructions. Installation of these seals is a critical part of this installation and, improperly made, will not provide the protection needed to isolate accidental explosions. Raceways in Class I hazardous locations are required to be provided with sealing fittings to prevent the passage of gases, vapors, or flames through the raceway from one portion of an electrical installation to another and to prevent pressure piling or cascading. Pressure piling can occur in these locations when an explosion in some part of the system, such as a switch enclosure, is not effectively blocked or sealed off. The first explosion can be made harmless if properly confined by a seal to the switch enclosure. If allowed to enter the raceway, it builds up great pressure, and another, more powerful explosion can follow. The second explosion builds up additional pressures in the gas-filled conduit, and a greater explosion results. This can continue in very long raceway runs not provided with intermittent seals until some part of the system ruptures. The hot or burning gases could then escape and ignite flammable gases in the surrounding atmosphere, causing a general explosion. Proper sealing is one of the most important requirements for safety in a Class I hazardous location. Explosive vapors are prevented from passing to a non-hazardous area when they could be ignited and carry the flame back to the explosive vapor location. In the manner of a pilot light on a gas range, arcs, sparks, and hot particles from an arcing device are also prevented from entering the raceway by the seal. Sealing fittings are designed to permit the pouring of an approved water-base sealing compound into the fitting around and between the conductors (see Figure 4–17). Conduit seals are limited to a 25% conductor fill by their listing. Where raceways are filled to 40% check with the manufacturer for conductor fill limits (see Figures 4–15 and 4–16).

The raceway should first be blocked at the bottom of the sealing fitting with a packed material to prevent the sealing mixture from draining out of the fitting. The liquid mixture solidifies after a few hours and becomes very hard. Proper mixing of the compound and liquid is essential; otherwise, the seal will not prevent the pressure piling. Even though the seal fitting might prevent the passage of vapors, arcs, and hot particles, all voids under, between, and over conductors must be filled if the sealing mixture is to perform its function of preventing the passage of gases, vapors, or flames through the raceway. The elimination of the pressure piling can be achieved by installing effective seals in every raceway regardless of size where it enters or leaves any junction box and at regular intervals. Seals should also be installed in all raceway runs. The minimum installation is to install a seal at all points in the system where they are required by the *NEC®*. In the absence of tests for various gases, to determine the proper spacing of seals, in long conduit runs it is not unreasonable to specify that they be installed at 50-foot intervals in all conduit runs within a Class I, Division 1 location.

If installed in accordance with the *NEC®* requirements, a rigid steel conduit system with suitable enclosure and seals will effectively contain an internal explosion and prevent the escape of burning gases until they are cooled sufficiently to prevent the ignition of flammable vapors in the surrounding atmosphere. The gases are cooled by expansion as they escape through the machine-grounded joints of predetermined width and through threaded joints. Ground joints, unless properly maintained, are not as reliable as threaded joints. A single loose screw in a cover of a ground joint, or scratches in ground services, will render an otherwise explosion-proof enclosure ineffective. Threaded joints, on the other hand, are reliable because the gases escape only by following the threads. The *NEC®* requires that at least five full threads be engaged at each coupling. Effective seals must be placed in a raceway as required by the *NEC®* within 18 inches of an arcing device or at or near the point where the raceway leaves or enters a hazardous area. In the case of a raceway 2 inches in diameter or larger, seals are required within 18 inches of any junction box containing splices, taps, or terminal connections. No coupling or fitting is permitted between the seal and the boundary between the hazardous and nonhazardous areas. In many cases, special panelboards and controls are designed with factory seals that are an integral part of the equipment designed for Class I locations. Because such seals effectively isolate the arc-producing segments of the assembly, additional seals are not generally required in the conduit leaving these panelboards. Incandescent and fluorescent fixtures in Class I locations are also available with integral seals (see Figure 4–18).

EYD
Drain seal
1 1/4 – 4 inches

EYS
Elbow seal
3/4-inch

EZD
With drain cover
1/2 – 2 inches

EZD
With inspection cover
1/2 – 2 inches

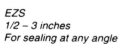

EZS
1/2 – 3 inches
For sealing at any angle

Figure 4–15 Examples of fittings that are approved for use in hazardous locations. (Courtesy of Crouse-Hinds, Division of Cooper Industries.)

Figure 4–16 Cutaway view of a new seal fitting that is sized for a 40 percent fill, the same as all typical raceways approved for installation in Class I locations. (Courtesy of Appleton Electric Company.)

Seals are not intended to prevent the passage of gases or vapors at a continuous pressure. Where there are differences in pressure across the seal, there can be some passage of gases or vapors. Leakage and the passage or propagation of flames through the interstices of stranded conductors can be reduced through the use of compact stranding or sealing of the individual stranding.

Although wiring methods are less restrictive in a Class I, Division 2 location, sealing is still required as specified in *NEC*® Section 501-5(b). In each conduit run passing from a Class I, Division 2 location into an unclassified area, a seal fitting is required:

- This sealing fitting can be located on either side of the boundary. It must be designed and installed to minimize the amount of gas or vapor that might enter the conduit system within the Class II location.

- It must also prevent the gas or vapor from being communicated to the conduit beyond the seal.

- Rigid metal conduit or threaded steel intermediate conduit must be used between the sealing fitting and the point at which the conduit leaves the Division 2 location.

- A threaded connection shall be used at the sealing fitting.

- Fittings or couplings are not permitted between the sealing fitting and the point at which the conduit leaves the Division 2 location.

- Conduit systems terminating to an open raceway in an outdoor unclassified area are not required to be sealed between the point at which the conduit leaves the classified location and enters the open raceway.

- The minimum requirements for the location and installation of sealing fittings are found in *NEC*® Section 501-5(a), (b), (c), (d), (e), and (f).

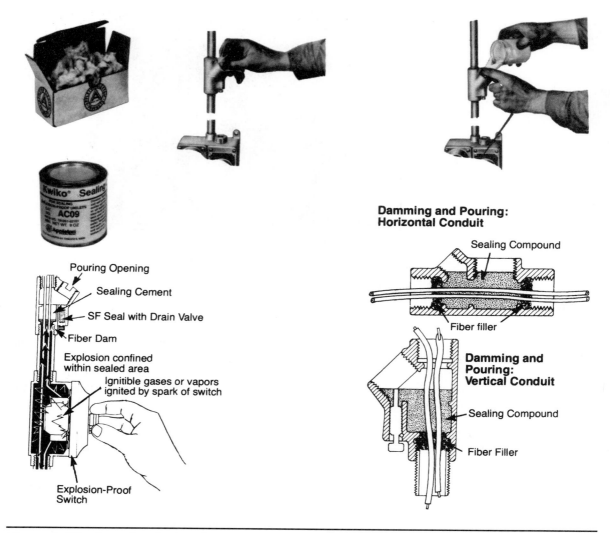

Figure 4–17 Examples of sealing fittings. Methods for damming and pouring sealing compounds. (Courtesy of Appleton Electric Company.)

Special type seals are available for retrofitting seals on existing installations. Cable fittings are available for the new type MC explosion-proof cable that will meet the new Exception 2 to Section 501-4(a) (see Figures 4–19 and 4–20).

Drainage

Where it is likely that water or other condensed vapors might be trapped in the system equipment enclosures or in the raceways, an approved means must be provided to prevent such accumulation or to provide a means for the drainage of those fluids or vapors periodically (see Figure 4–21). An integral means for periodic drainage shall be provided by the

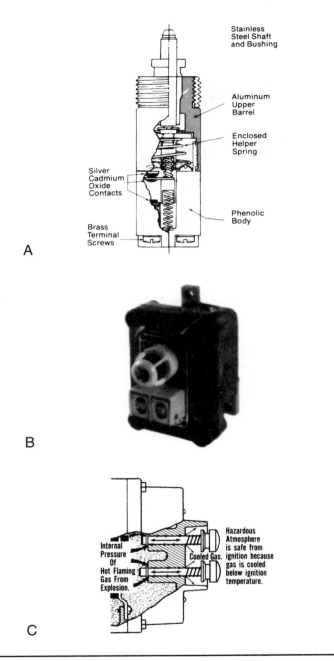

Figure 4–18 Examples of factory-sealed (A) pilot light and (B, C) push-button stations. (A and B courtesy of Appleton Electric Company; C courtesy of Crouse-Hinds, Division of Cooper Industries.)

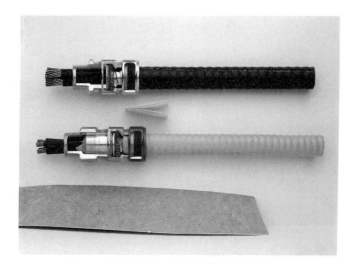

Figure 4–19 Examples of fitting for type MC cable permitted in accordance with *NEC®* Articles 501 and 502. (Courtesy of Appleton Electric Company.)

Figure 4–20 Example of a new type split-seal fitting that can be used in existing Division 2 locations when needed.

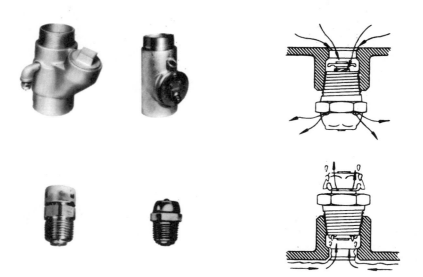

Figure 4–21 Example of seals and drain and vent fittings. (Courtesy of Appleton Electric Company.)

manufacturer of motors and generators where needed, or the conduit system and equipment shall be arranged to minimize the entrance of liquids and vapors acceptable to the authority having jurisdiction. Equipment, such as canned pumps or process or service connections for flow, pressure, analysis measurement, and so on that depend on a tube, single compression seal, or diaphragm to prevent flammable or combustible fluids from entering the raceway system must also have an additional approved seal, barrier, or other means to prevent propagation if the primary seal should fail. The additional redundant means must be approved for the temperature and pressures that they might be subjected to if a primary means should fail.

Drains, vents, or other devices are required so that the primary seal leakage will be obvious.

CHAPTER 5

Class II

Class II Locations

Class II locations are those that are hazardous because of the presence of combustible dust.

Class II Location Examples

- Grain elevators and bulk handling facilities
- Magnesium manufacture and storage
- Starch manufacture and storage
- Fireworks manufacture and storage
- Flour and feed mills
- Areas for packaging and handling of pulverized sugar and cocoa
- Facilities for the manufacture of magnesium and aluminum powder
- Some coal preparation plants and coal handling facilities
- Spice grinding plants
- Confectionery manufacturing plants (see Figures 5–1 and 5–2.)

See NFPA 499 for the classification of combustible dusts and harzardous (classified) locations for electrical installations in chemical process areas. This document was newly developed in 1997, combining part of NFPA 497M and NFPA 497B. It provides typical calculations for area classification and an expanded appendix. **Note:** When the division is not specified, the equipment is suitable for both Division 1 and 2 applications. **Warning:** Where Class II, Group E dusts are present in hazardous quantities, this area (or areas) is classified Class I, Division 1. There are no Division 2 areas. Dusts containing magnesium or aluminum dust are especially hazardous; *extreme caution* is advised to avoid ignition and explosion.

Class II Group Classifications

Dusts have also been placed in Groups E, F, and G on the basis of their particular hazardous characteristics and the dusts' electrical resistivity. It is important to select equipment suitable for the specific hazardous group to ensure that the installation is as safe as possible.

Figure 5–1   Class II grain elevator locations. (Courtesy of Appleton Electric Company.)

Figure 5–2   Grain storage silos can be extremely dangerous. It may be possible to avoid disasters like this by eliminating common sources of ignition. (Courtesy of Appleton Electric Company.)

Class II, Division 1

Class II locations are areas where the combustible dust might be in suspension in the air under normal conditions and in sufficient quantities to produce explosive or ignitable mixtures. The classification of the area would apply where the combustible dusts are likely to be present continuously, intermittently, or periodically. Division 1 locations also exist where failure or malfunction of machinery or equipment might cause a hazardous location to exist while providing a source of ignition with the simultaneous failure of electrical equipment. Included also are locations in which combustible dust of an electrically conductive nature might be present.

Class II, Division 1 Location Examples

- Areas used for the processing and handling of grain products
- Areas used for the processing and handling of cocoa, sugar, egg, and milk powders

- Areas used for the processing and handling of pulverized spices, starch, and paste
- Areas used for the processing and handling of potato and wood flour
- Areas used for the processing and handling of oil meal from seeds or beans
- Areas used for the processing and handling of dusts containing aluminum or magnesium

Class II, Division 2

A Class II, Division 2 location is one in which combustible dust will not normally be in suspension in the air and normal operations will not put the dust in suspension in the air but where accumulation of the dust might interfere with the safe dissipation of heat from electrical equipment or where accumulations near electrical equipment might be ignited by arcs, sparks, or burning material from the equipment.

Class II, Division 2 Examples

- Grain elevators and bulk handling facilities
- Magnesium storage
- Starch storage
- Fireworks storage
- Areas within flour and feed mills
- Areas for storing pulverized sugar and cocoa
- Aluminum powder storage areas
- Areas in some coal preparation plants and coal handling facilities
- Areas for storing spices in grinding plants
- Storage areas in confectionery manufacturing plants

Note: The authority having jurisdiction might determine that areas can be classified Division 2 or considered nonhazardous areas if adequate dust removal systems are installed and maintained or where products such as seed are handled in a manner that assures limitation of dust quantities to warrant declassification to a lower class or in some cases to an unclassified area. For guidance, see NFPA 499.

Equipment for Class II Locations

The enclosures used to house devices in Class II locations are designed to seal out dust. Contact between the hazardous atmosphere and the source of ignition has been eliminated, and no explosion can occur within the enclsoure (see Figures 5–3 and 5–4). Equipment and wiring of the type defined in Article 100 as explosion-proof is not required and is not acceptable in Class II locations unless approved for such locations.

As in Class I equipment, heat-producing equipment must be designed to operate below the ignition temperature of the hazardous atmosphere. However, in Class II equipment, additional consideration must be given to the heat buildup that can result from the layer of dust that will settle on the equipment.

Specific Equipment for Class II Locations

Class II, Division 1 Locations

Switches, circuit breakers, motor controllers, and fuses, including push-buttons, relays, and similar devices, that interrupt current during normal operation or are installed where electrically conductive combustible dusts might be present must be installed in

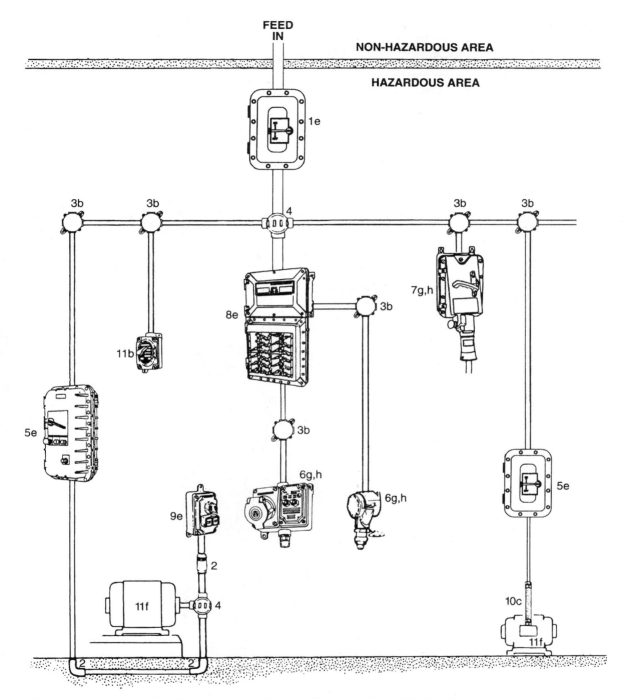

FEED IN

NON-HAZARDOUS AREA

HAZARDOUS AREA

1e

3b   3b   4   3b   3b

7g,h

8e   3b

11b

5e

3b

6g,h

9e   6g,h

5e

2

11f   4

10c

11f

*Where Class II, Group E dusts are present in hazardous quantities, there are only Div. 1 locations.

Figure 5–3   Power diagram for Class II, Division 1 and 2 locations. (Courtesy of Appleton Electric Company.)

Key to Product

1—Circuit Breaker or Disconnect Switch. EB, EDS series
2—Unions/Elbows. UNY-NR, UNF-NR, UNY/F UNL, UNYL/ UNFL; ELF, ELMF.
3—Junction Box. CPU, with ground surface cover.
4—Junction Box. GR, with screw cover.
5—Combination/Motor Starter. E Series, EXMS.
6—Receptacle. EFS U-Line with GFI, CPS, FSQX.
7—Receptacle, Interlocked. DBR, EBR, JBR.
8—Panelboard. EWP, D2P, GB1, FB2.
9—Push Button Station. EFDB, N2, EFS, EFD, UniCode,
10—Flexible Conduit, liquidtight, ST Fittings.
11—Motor for Location.

National Electric Code® Reference

a—*Section 502-4(a).* Wiring methods shall be threaded rigid metal conduit, steel IMC or MI cable with approved fittings. MC cable with approved fittings permitted.

b—*Section 502-4(1).* Fittings and boxes must have threaded bosses, tight-fitting covers, and no holes for dust to enter.

c—*Section 502-4(2).* Explosionproof flexible connections NOT required; dusttight flexible conduit or hard usage cord permitted.

d—*Section 502-5.* Sealing fittings in Class II Div. 1 and 2 need not be explosionproof.

e—*Section 502-6(1).* In Div.1, equipment to interrupt current shall have approved dust-ignitionproof enclosures. In Div. 2 enclosures should be dusttight.

f—*Section 502-8.* Motors and generators in Class II, Division 1 need to be approved. For Division 2 see table 500-3(f).

g—*Section 502-13.* In Division 1, receptacles and attachment plugs shall be approved for Class II. In Division 2, connection to the supply circuit cannot be made or broken while live parts are exposed. They are not required to be Class II approved.

h—Plugs and receptacles are not suitable for Group E.

Figure 5–3 *Continued.*

Figure 5–4 Example of a Class II switched receptacle. (Courtesy of Crouse-Hinds, Division of Cooper Industries.)

dust-ignition-proof enclosures. The enclosed equipment must be listed or approved as a complete assembly for Class II locations.

In areas where electrical combustible dusts are not present, switches containing no fuses and not designed or intended to interrupt current can be enclosed within metal enclosures that shall be designed to minimize the entrance of dust and that shall be equipped with telescoping or close-fitting covers or with other acceptable means to prevent the escape of sparks and burning material or to allow the entrance of combustible dusts (see Figure 5–6). The shall have no openings (i.e., holes for attachment screws).

In locations where dust from magnesium, aluminum, aluminum bronze powders, or other metals of similarly hazardous characteristics might be present, fuses, switches, motor controllers, and circuit breakers shall have enclosures specifically approved for such locations.

Listed or approved dust-ignition-proof enclosures are required for control transformers, solenoids, impedance coils, resistors, and any overcurrent devices or associated switching mechanisms (see Figure 5–7). This equipment shall not be installed in a location where dust from magnesium, aluminum, and other similar powders are present, unless they are installed within an enclosure approved specifically for the purpose.

Aluminum bronze powders or other similarly hazardous metal dusts must not be present unless provided with an enclosure approved for the specific location. Motors, generators, and other rotating electrical machinery located in Division 1 locations shall be approved for the location or must be of a totally enclosed pipe-ventilated type in accordance with Sections 502-9 and 502-9(a) and must meet the temperature requirements in accordance with Section 502–1.

Lighting fixtures for fixed and portable lighting must be approved for the location and marked to indicate the maximum lamp wattage permitted. Where dust from magnesium, aluminum, aluminum bronze powders, or metals with similar characteristics might be present, the fixtures, both fixed or portable and all auxiliary equipment, must be approved for the location. Each fixture is to be protected from physical damage by a guard or be located where not subject to damage (see Figures 5–5 and 5–8).

Pendant and chain-hung fixtures must be installed with an approved means and comply with the requirements of Section 502-11(a)(3) and supported in accordance with Section 502-10(a).

All utilization equipment shall be approved for Class II locations. In addition, when dust from magneisum, aluminum, aluminum bronze powders, or other metals of similarly hazardous characteristics might be present, such equipment shall be approved for the specific location.

Flexible cords used in Class II, Divisions 1 and 2 must comply with Section 502-12. They must be approved for extra-hard usage and supported by suitable means to prevent tension on the terminations, contain a grounding conductor complying with Section 400-23, and be spliced or terminated in an approved manner. Suitable seals shall be installed to prevent the entrance of dust where the flexible cord enters boxes or fittings that are required to be dust-ignition-proof.

Receptables and attachment plugs shall be designed to provide connections for the grounding conductor of the flexible cord and shall be approved for Class II locations. **(FPN):** See *NEC*® Article 800 for rules governing the installation of communication circuits.

Signaling, alarm, remote control, and communication systems, as well as meters, instruments, and relays, shall comply with the following.

The wiring methods must be in accordance with Section 502-4(a). Enclosures approved for a Class II location are required for switches, circuit breakers, relays, contactors, fuses,

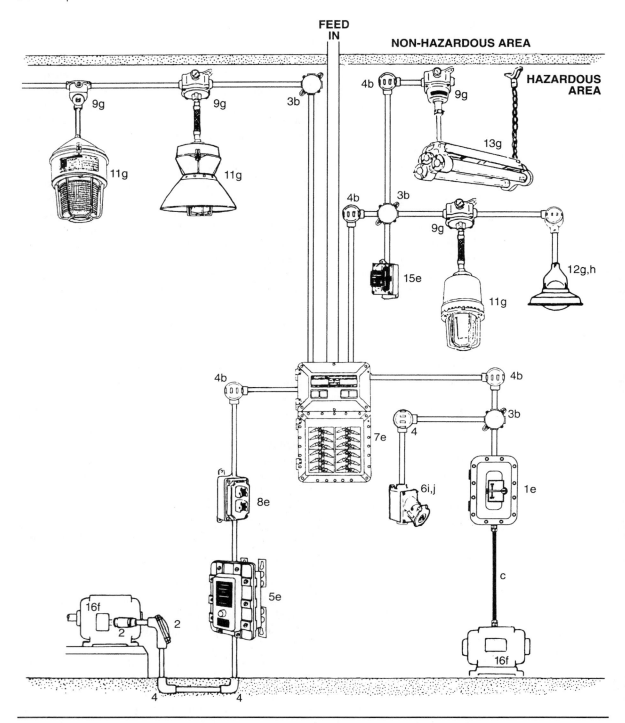

Figure 5–5 Lighting diagram for Class II, Division 1 and 2 locations. (Courtesy of Appleton Electric Company.)

Key to Product

1—Circuit Breaker, Disconnect Switch, Manual Starter. EB, EDS, EXMS.

2—Unions/Elbows, UNY-NR, UNF-NR, UNY/F UNL, UNYL/UNFL, ELF, ELMF.

3—Junction Box. CPU, ELBD, with ground surface cover.

4—Junction Box. GR, with screw cover.

5—Manual Motor Starter. EXMS.

6—Receptacle. EFS U-Line.

7—Panelboard, EWP, D2P, GB1, FB2, EPB.

8—Push Button/Pilot Light, EFDB, N2, EFS, EFD, UniCode.

9—Fixture Hangers. Div. 1—CPU, GRF, EFHC, EFHU, EXJF.

10—Fixture Hangers. Div. 2—JB cushion, GS cushion, AHG cushion.

11—Fixtures, Div 1, HID—Codemaster, Codemaster Jr., Mercmaster III, Mercmaster II.

12—Fixtures, Div. 1, Incandescent—A-51, DT, EDTP, EHL.

13—Fixtures, Div. 1, Fluorescent—EFU, MMIII Low Profile, APL, PAPL.

14—Fixtures, Div. 2 (see explanation "h" opposite).

15—Switches. EFS, EDS, Contender.

16—Motors for Location.

* Where Class II, Group E dusts are present in hazardous quantities, there are only Div. 1 locations.

National Electric Code® Reference

a—*Section 502-4(a)*. Wiring methods shall be threaded rigid metal conduit, steel IMC or MI cable with approved fittings. Approved MC cable with approved fittings permitted.

b—*Section 502-4(1)*. Fittings and boxes must have threaded bosses, tight-fitting covers, and no holes for dust to enter.

c—*Section 502-4(2)*. Explosionproof flexible connections NOT required, dusttight flexible conduit or hard usage cord permitted.

d—*Section 502-5*. Sealing fittings in Class II Div. 1 and 2 need not be explosionproof.

e—*Section 502-6(1)*. Equipment to interrupt current shall have approved dust-ignitionproof enclosures.

f—*Section 502-8*. Motors and generators in Division 1 shall be approved. For Division 2 see table 500-3(f).

g—*Section 502-11*. In Division 1, fixtures shall be approved for Class II. Pendant fixtures with stems longer than 12 inches shall be braced or have an approved flexible connector.

h—*Section 502-11(b)(2)*. In Division 2, if not approved for Class II, fixtures shall be designed to minimize dust deposits on lamps and prevent escape of sparks. They must be marked to show maximum lamp wattage per 500-3(f)

i—*Section 502-13*. In Division 1, receptacles and attachment plugs shall be approved for Class II. In Division 2, connection to the supply circuit cannot be made or broken while live parts are exposed. They are not required to be Class II approved.

j—Plugs and receptacles are not suitable for Group E.

Figure 5–5 *Continued.*

Figure 5–6 Example of a conduit box suitable for use in Class II locations. (Courtesy of Crouse-Hinds, Division of Cooper Industries.)

**Typical Explosionproof
Electrical Controls.**

Figure 5–7   Examples of enclosures suitable for
use in accordance with Section 502-6(a)(2).
(Courtesy of Crouse-Hinds, Division of Cooper
Industries.)

Figure 5–8   Examples of lighting fixtures suitable
for use in Class II locations. (Courtesy of Crouse-
Hinds, Division of Cooper Industries.)

and current-breaking contacts for bells, horns, howlers, sirens, and other devices that might
produce sparks or arcs (see Figure 5–9). Resistors, transformers, and other heat-generating
equipment shall be provided with enclosures approved for Class II locations. General purpose enclosures are permitted where current-breaking contacts are immersed in oil or the
interruption of current occurs within a chamber sealed against the entrance of dust.

Motors, generators, and other rotating electric machinery must meet the requirements
of Section 502-8(a).

All wiring and equipment shall be approved for Class II locations associated with dusts
that are electrically conductive. Where dust from magnesium, aluminum, aluminum bronze
powders, or other metals of similarly hazardous characteristics might be present, all apparatus and equipment shall be approved for the specific conditions (see Figure 5–10). *NEC*®
502-15 states, "Live parts shall not be exposed."

Signaling, Alarm, Remote-Control, and Communication Systems.

ETH Flex-Tone Signaling Device

ETW Telephone

*EFS
Pushbutton Station and pilot light*

ESR Bell

ETH Horn signal

EV Strobe warning light

Figure 5–9
Examples of signaling, alarm, remote control, and communications devices suitable for use in Class II locations. (Courtesy of Crouse-Hinds, Division of Cooper Industries.)

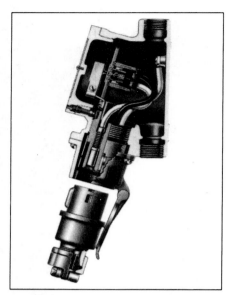

Receptacle constructed with an interlocked switch. Rotating the plug after insertion actuates this switch. This is also referred to as "dead front."

Figure 5–10 ARE receptable cutaway with a spring door. (Courtesy of Crouse-Hinds, Division of Cooper Industries.)

Division 1 Grounding Requirements

Wiring and equipment must comply with the provisions of Article 250. Locknut bushings and double-locknut types of contact are not acceptable for bonding purposes. Bonding jumpers or other approved means must be used to bond all intervening raceways, fittings, boxes, enclosures, between Class II locations and the point of grounding for service equipment or separately derived systems, except as permitted in Section 502-16(a) and (b) and the exception. Where flexible conduit is permitted by Section 502-4), parallel internal or external bonding jumpers complying with Section 250-102 shall be installed. Additional applicable bonding requirements for hazardous (classified) locations are found in Section 250-100.

Surge arresters are designed for specific duty cycle and must comply with Article 280 and, in addition, shall be in suitable enclosures. Surge-protective capacitors must be designed for the specific duty in accordance with Article 460.

A separate grounded conductor shall be installed in each single-phase branch circuit that is part of a multiwire branch circuit except where the disconnect device(s) opens all ungrounded conductors simultaneously.

Class II, Division 2 Locations

Enclosures for fuses, switches, circuit breakers, and motor controllers, including pushbuttons, relays, and similar devices, shall be dust-tight (see Figure 5–10). Control transformers, solenoids, impedance coils, resistors, and resistance devices with associated equipment must be installed in dust-tight enclosures. Control transformers, solenoids, and impedance coils without switching or arcing devices can be installed in tight metal housings without ventilating openings.

Electrically heated utilization equipment is generally required to be approved for Class II locations; see Section 502-10(b)(1) and exception. Motors and motor-driven utilization equipment are required to comply with Section 502-8(b) and exception.

Portable lighting equipment must be clearly marked to indicate maximum wattage and be approved for the location. Fixed lighting must be approved for Class II locations or shall provide enclosures for the lamps and lampholders that are designed to prevent the escape of sparks, burning material, or hot metal and to minimize dust from accumulating on the lamps. Each fixture shall be marked with the maximum allowable wattage and protected from physical damage by suitable guards or by location. Pendant and chain-hung fixtures shall comply with Section 502-10(b). Starting and control equipment for electric-discharge lamps shall comply with the requirements of Section 502-7(b).

Flexible cords in Class II, Divisions 1 and 2 are required to comply with Section 502-12. All flexible cords must be of a type approved for extra-hard usage with an equipment-grounding conductor. Terminations must be made in an approved manner, and the cord must be supported or fastened in a manner so that there will be no tension on the terminals. Receptacles and attachment plugs shall provide connections for the grounding conductor of the flexible cord and shall be designed so that connection to the supply circuit cannot be made or broken while live parts are exposed.

Signaling, alarm, remote control, and communication systems, as well as meters, instruments, and relays, must comply with the following.

Arc-producing contacts shall be enclosed in approved enclosures in accordance with Section 502-14(a) or shall have tight metal enclosures designed to minimize the entrance of dust with no openings through which, after installation, sparks or burning material might escape. Nonincendive circuits that under normal conditions do not release sufficient energy to ignite a dust layer are permitted in general purpose enclosures.

The windings and terminal connections of transformers and similar equipment shall be provided with tight metal enclosures without ventilating openings. Resistors and similar equipment shall comply with Section 502-14(a)(3). Where the maximum operating temperature will not exceed 120°C (248°F), general purpose enclosures are permitted for thermionic tubes, nonadjustable resistors, or rectifiers.

Motors, generators, and other rotating electric machinery shall comply with Section 502-8(b) and the wiring method with Section 502-4(b). *NEC*® Section 502-15 states, "Live parts shall not be exposed."

Division 2 Grounding Requirements

Wiring and equipment in Class II Division 2 locations shall be bonded as specified in Article 250 generally, and in Sections 250-100 and 502-16(a) specifically. Flexible conduit used as an equipment-grounding conductor in accordance with Section 502-4 shall be installed with internal or external bonding jumpers complying with Section 250-102.

The bonding jumper can be deleted where listed liquid-tight flexible metal is used and the conduit is 6 feet (1.83 m) or less in length, with listed fittings, having the overcurrent protection at 10 amperes or less, and the load is not a power utilization load.

Surge arresters, including their installation and connection, shall comply with Article 280. Surge-protective capacitors shall be of a type designed for specific duty.

Wiring Methods for Class II Locations

Class II, Division 1 Wiring Methods

Class II, Division 1 locations permit threaded rigid metal conduit, threaded steel intermediate metal conduit or type MI cable with termination fittings approved for the location and installed to avoid tensile stress at the fittings. The fittings and boxes with threaded

Figure 5–11 An example of a Class II panelboard. (Courtesy of Crouse-Hinds, Division of Cooper Industries.)

bosses and close-fitting covers shall have no openings through which dust might enter or sparks or burning material might escape (see Figure 5–11). In locations where combustible dusts of an electrically conductive nature are likely, the fittings and boxes that are designed to contain taps, joints, or terminal connections must be approved for Class II locations. Flexible connections shall be made with dust-tight flexible connectors, liquid-tight flexible metal conduit with approved fittings, liquid-tight flexible nonmetallic conduit with approved fittings, or flexible cord approved for extra-hard usage and provided with bushed fittings. Flexible cords must comply with Section 502-12. Flexible connections where subject to oil or other corrosive conditions shall be of a type approved for the condition or shall be protected by a suitable sheath. **(FPN):** See Section 502-16(b) for grounding requirements where flexible conduit is used.

In industrial establishments with restricted public access, and where conditions of maintenance and supervision assure that only qualified persons will service the installation, type MC (*NEC*® Article 334) may be used, provided that the cable type is listed for the use in Class II, Division 1 locations. The cable must have a gas- and vapor-tight continuous corrugated aluminum sheath and an overall jacket of suitable polymeric material. The termination fittings must be listed for the use. The type MC cable must also contain an equipment-grounding conductor sized in accordance with *NEC*® Section 250-122 (see Figure 4–18).

Class II, Division 2 Wiring Methods

Class II, Division 2 locations permit the use of rigid metal conduit, intermediate metal conduit, electrical metallic tubing, dust-tight wireways, or types MI or MC cable with approved termination fittings or types MC, PLTC, or TC cable installed in ventilated channel-type cable trays in a single layer with a space of not less than the larger cable diameter between the two adjacent cables. Using any of the methods suitable for wiring in ordinary locations shall be permitted, except wiring that under normal conditions cannot release sufficient energy to ignite a specific combustible dust mixture by opening, shorting, or grounding. Nonincendive circuits are permitted in wiring methods suitable for ordinary locations.

Sealing Requirements for Class II, Divisions 1 and 2

Combustible and explosive dusts constitute a hazard in Class II locations. Such dusts do not travel for great distances within a conduit. Therefore, seals are not required if the conduit running between a dust-ignition-proof enclosure and enclosure that is not dust-ignition-proof consists of either a horizontal section not less than 10 feet in length or a vertical section not less than 5 feet in length extending downward from the dust-ignition-proof enclosure through the raceway. Where such conduit runs are not provided, seals must be used to prevent the entrance of dust into ignition-proof enclosures through the raceway. Sealing fittings and poured sealing compound can be used in both Division 1 and Division 2, and they are not required to meet the same stringent requirements as those in Section 501-5. The term "dust-ignition-proof" is defined in *NEC*® Section 502-1.

CHAPTER 6

Class III

Class III Locations

Class III locations are those that are hazardous because of the presence of easily ignitable fibers or flyings. It includes all locations in which the fibers or flyings are handled, manufactured, used, or stored. It is often believed that Class III hazards are far less hazardous than are Class I and Class II areas. This classification deals with different materials and a different type of hazard. However, they are dangerous not only because of ignitability but also because of progapation, the speed at which the fire with spread after it is ignited. When these fibrous materials, such as cotton lint that has settled on equipment, machinery, and building framing members, catch on fire, the fire can and very often does spread at speeds approaching multiple explosions. These fires are called "flash fires" and have been the origins of some of our country's greatest disasters.

Class III, Division 1 Locations

These locations are generally locations where the hazardous (classified) materials are manufactured, handled, and used.

Class III Location Examples (Where Division 1 and Division 2 Might Be Found)

- Textile mills (i.e., cotton and rayon)
- Cotton gins
- Cotton seed mills
- Woodworking plants
- Combustible fiber manufacturing plants (i.e., hemp, cocoa, oakum, excelsior, cotton linters, and kapok). (See Figures 6–1 and 6–2.)

Class III, Division 2 Locations

These locations are generally areas or structures where easily ignitable fibers are stored or handled. Examples of locations where Division 2 areas might be are listed above.

Figure 6–1 Example of a Class III manufacturing area. (Courtesy of Appleton Electric Company.)

Figure 6–2 Example of a Class III storage area. (Courtesy of Appleton Electric Company.)

Equipment for Class III Locations

Class III locations require equipment that is designed to prevent the entrance of fibers and flyings and to prevent the escape of sparks or burning material. Equipment installed in these locations must be capable of operating at its full load rating without developing surface temperatures hot enough to cause excessive dehydration or gradual carbonization of accumulated fibers or flyings (i.e., lint). Materials that are very susceptible to spontaneous ignition carbonize when they are excessively dry. Surface temperatures shall not exceed 165°C (329°F) for equipment not subject to overloading. Equipment, such as motors or transformers, that might be overloaded are limited to a maximum operating temperature of 120°C (248°F). Transformers and capacitors located in Class III, Divisions 1 and 2 must comply with the Class II requirements in Section 502-2(b).

Specific Equipment for Class III Locations

Class III, Division 1 and 2 Locations

Dust-tight enclosures are required for switches and motor controllers, including push-buttons, relays, similar devices, and overcurrent protection.

Dust-tight enclosures meeting the temperature requirements in Section 503-1 are required for heat-producing equipment such as transformers and resistors used for the control of motors, appliances, and other electrical equipment.

Only totally enclosed nonventilated, totally enclosed pipe-ventilated, or totally enclosed fan-cooled motors, generators, and other rotating machinery are permitted unless otherwise allowed by the authority having jurisdiction. **Note:** Guidance to help the AHJ is found in the Section 503-6 exception.

Ventilating pipes must comply with Section 503-7 and shall be constructed of metal 0.021 inch (533 micrometers) in minimum thickness or of equally substantial noncombustible material and shall be protected against physical damage and corrosive influences. The piping shall be routed directly to a source of clean air outside with screened covers to prevent the entrance of small animals or birds.

Utilization equipment such as electrically heated utilization equipment shall be approved for Class III locations. Motors of motor-driven utilization equipment must comply with Section 503-6. Switches, circuit breakers, motor controllers, and fuses must comply with Section 503-4.

Only enclosures for lamps and lampholders that minimize the entrance of fibers and flyings and prevent the escape of sparks, burning material, or hot metal are permitted for

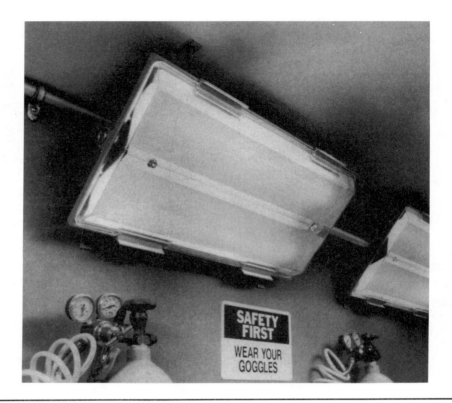

Figure 6–3 Example of a compact fluorescent light fixture suitable for use in Class III locations. (Courtesy of Appleton Electric Company.)

fixed lighting. Fixtures shall not be subject to physical damage or shall be protected with suitable guards and be marked for the maximum permitted wattage. The surface temperature shall comply with Section 503-1 (see Figure 6–3). Pendant fixtures shall comply with Section 503-9(c). Portable lighting equipment must comply with the provisions of Section 503-9(d).

Flexible cords shall be approved for extra-hard usage, contain a grounding conductor in accordance with Section 400-23, and shall be supported with an approved means to prevent tension on the terminations. Fittings shall be designed to prevent the entrance of fibers or flyings where the cord enters boxes, and all terminations shall be made in an approved manner.

Only receptacles and attachment plugs of the grounding type designed to minimize the accumulation or the entry of fibers or flyings and prevent the escape of sparks or molten particles are permitted unless otherwise permitted by the AHJ.

Signaling and intercommunication systems shall comply with the requirements of Article 503 regarding wiring methods, equipment, and related components.

Where installed for operation over combustible fibers or accumulations of flyings, traveling cranes and hoists for material handling, traveling cleaners for textile machinery, and similar equipment shall comply with the provisions of Section 503-13(a) through (d).

Storage-battery charging equipment shall be located in well-ventilated rooms to exclude flyings or lint and be built or lined with noncombustible materials. Live parts shall not be exposed, except as permitted in Section 503-13.

Figure 6–4   Example of Class III conduit bodies. (Courtesy of Crouse-Hinds, Division of Cooper Industries.)

Wiring Methods

Class III, Division 1 Locations

The permitted wiring methods are rigid metal conduit, intermediate metal conduit (IMC), electrical metallic tubing (EMT), rigid nonmetallic conduit, and type MC or MI cable with approved termination fittings or dust-right wireways. Flexible connections must be made with dust-tight flexible fittings and installed in accordance with Section 503-16)(b), using liquid-tight flexible metal conduit with approved fittings, liquid-tight flexible nonmetallic conduit with approved fittings, or flexible cord in conformance with Section 503-10 (see Figure 6–4). Boxes and fittings must be dust-tight.

Class III, Division 2 Locations

Wiring methods approved for Class III, Division 1 locations are acceptable. In addition, in compartments or locations solely for storage and containing no machinery, the use of open wiring on insulators can be used in accordance with all the provisions of Article 320 when not subject to physical damage.

Grounding and Bonding Class III, Divisions 1 and 2

Class III wiring and equipment must comply with Article 250 (see Section 503-16). Bonding jumpers with proper fittings or other approved means of bonding shall be used and apply to all intervening raceways, fittings, boxes, enclosures, and so on between Class III locations and the point of grounding for service equipment or the point of grounding of a separately derived system except as otherwise permitted in the Section 503-15 exception. Locknut-bushing and double-locknut types shall not be depended on for bonding Section 503-16(a).

Equipment grounding conductors shall be installed in accordance with Section 250-118 and 503-3. Where flexible conduit is used, it must be installed with internal or external bonding jumpers in parallel with each conduit and comply with Section 250-102 or in accordance with Section 503-15(b). Exception where all the conditions are met.

Sealing Requirements

Seals are not required in Class III locations.

CHAPTER 7

Specific Hazardous (Classified) Locations

Article 510 states that Articles 511 through 517 cover occupancies or parts of occupancies that are or might be hazardous because of atmospheric concentrations of gases, vapors, liquids, or materials that might be readily ignitable.

Electrical equipment and electrical wiring in occupancies within the scope of Articles 511 through 517 apply to the general rules of Chapters 1 through 4 except as modified by these articles. Where judged that unusual conditions exist, the authority having jurisdiction has the authority and can rule as to the application of the specific rules.

Commercial Garages: Repair and Storage

Commercial garages are defined by *NEC*® Article 511. Additional information can be obtained in NFPA 88A-1995 for parking structures and NFPA 88B-1997 for repair garages.

Scope: Article 511, Section 511-1: "These occupancies shall include locations used for service and repair operations in connection with self-propelled vehicles (including, but not limited to, passenger automobiles, buses, trucks, tractors, etc.) in which volatile flammable liquids are used for fuel or power" (Figure 7–1). It is important to bear in mind that similar commercial repair shops for such equipment as boat repair shops, lawnmower repair shops, and other shops for self-propelled vehicles used to store or repair volatile fuel-driven or propelled equipment, are now included in the scope of Article 511. These gasoline engine repair shops or areas within these shops can be hazardous (classified) locations and as such must also comply with the appropriate provisions of Article 501. Areas in which flammable fuel is transferred to vehicle fuel tanks must conform with *NEC*® Article 514 (see next page).

Parking garages that are used for parking or storage and where no repair work is done except the exchange of parts and routine maintenance requiring no use of electrical equipment, open flame, welding, or the use of volatile flammable liquids are not required to be classified but they shall be adequately vented with the capacity to carry the exhaust fumes of the engines.

Figure 7–1 Lubrication pit area for a commercial garage. (Courtesy of Appleton Electric Company.)

Table 514-2 (Partial Extract to Cover This Specific Requirement)

Lubrication or Service Room—

Class 1 Division 2

Entire area within any pit used without dispensing for lubrication or similar services where Class I liquids may be released.

Area up to 18 inches above any such pit, and extending a distance of 3 feet horizontally from any edge of the pit.

Entire unventilated area within any pit, below-grade area, or subfloor area.

Area up to 18 inches above any such unventilated pit, below-grade work area, or subfloor work area and extending a distance of 3 feet horixontally from the edge of any such pit, below-grade work area, or subfloor work area.

Nonclassified

Any pit, below-grade work area, classified or sub-floor work area that is ventilated in accordance with 5-1.3. NFPA 30A 1996

Article 500 Areas within 511 Structures

Class I, Division 2 apply to areas up to a level of 18 inches above each floor. The entire area up to a level of 18 inches above the floor is considered that classified location, except where the enforcing agency judges there is mechanical ventilation providing a minimum of four air changes per hour.

Class I, Division 1 shall apply to any pit or depression below floor level and to the area up to floor level. Where ventilation is provided in which six air changes per hour are exhausted at the floor level of the pit, the area can be classified as Class I, Division 2 by the enforcing agency (see Figure 7–2).

A change to the 1996 *NEC*® brought about a special set of rules for lubrication and service rooms. This was because of the limited amounts gasoline and cleaning solvents present in these normally franchised facilities. Although they clearly are commercial garages, a new exception was added to Section 511-3(b) that assigns these facilities to Article 514. The Section 511-3(b) exception states "Lubrication and service rooms without dispensing shall be classified in accordance with Table 514-2."

Areas not classified are areas separated and adjacent to classified locations. They are specifically defined and are areas or rooms in which flammable vapors are not likely to be released or to accumulate, for example, stock rooms, switchboard rooms, and similar locations where mechanically ventilated at a rate of four or more air changes per hour or where effectively cut off by walls or partitions.

The authority having jurisdiction can deem that in specific adjacent areas where no ignition hazard exists the area not require classification. The adequate ventilation, air pressure differentials, or physical spacing are valid reasons for making this determination.

Where volatile fuel-dispensing units are located within buildings, the requirements of Article 514 apply. Where mechanical ventilation is provided in the dispensing area, the controls shall be interlocked so that the fuel dispenser cannot operate without ventilation,

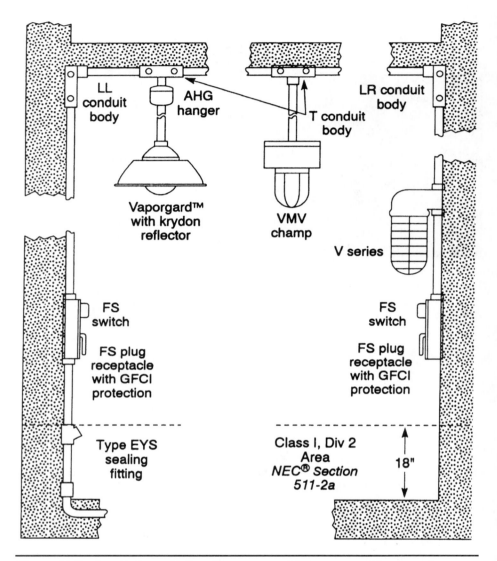

Figure 7–2   Installation guidelines for a commercial garage area. (Courtesy of Appleton Electric Company.)

as prescribed in Section 500-5(b). Liquid petroleum gas dispensers are prohibited within Article 511 structures.

All portable lighting equipment shall be approved for Class I, Division 1 unless the lamp and its cord are supported or arranged in such a manner that they cannot be used in the Section 511-3 locations (see Figure 7–3).

Wiring and Equipment

Wiring and equipment located within the Class I locations shall conform to the applicable provisions of Article 501. For raceways embedded in masonry walls or buried beneath the

Figure 7–3 Example of a portable floodlight suitable for use in hazardous locations; for other examples of portable lighting, see Figure 4–10 in Chapter 4. (Courtesy of Appleton Electric Company.)

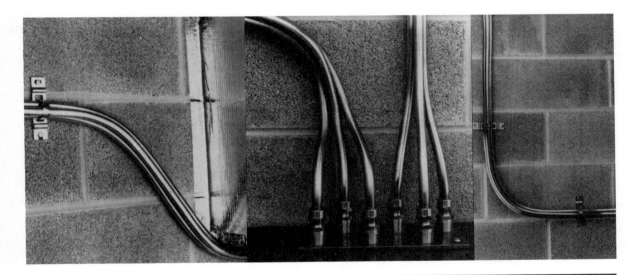

Figure 7–4 MI cable installation (Article 330). (Courtesy of Pyrotenax USA Inc., photo by Robert Stewart Associates.)

floor and if any connections or extensions lead into or through the area, seals are required as per Section 501-5 and 501-5(b)(2) and shall apply to both horizontal and vertical boundaries of the defined Class I location (see Figures 7–4 and 7–5; see Figure 7–1).

Exception No. 2: Rigid nonmetallic conduit complying with Article 347 is permitted where buried under not less than 2 feet (610 mm) of cover. Where rigid nonmetallic conduit is used, threaded rigid metal conduit or threaded steel intermediate metal conduit shall be used for the last 2 feet (610 mm) of the underground run to emergence or to the point of connection to the above-ground raceway, and an equipment grounding conductor must be included to provide electrical continuity of the raceway system and for grounding of the non-current-carrying metal parts.

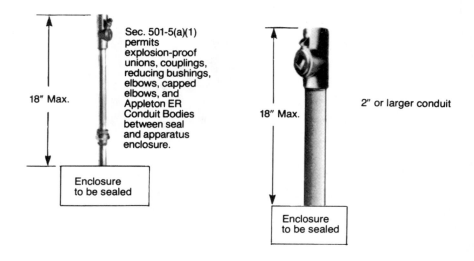

18″ Max.

Sec. 501-5(a)(1) permits explosion-proof unions, couplings, reducing bushings, elbows, capped elbows, and Appleton ER Conduit Bodies between seal and apparatus enclosure.

Enclosure to be sealed

18″ Max.

2″ or larger conduit

Enclosure to be sealed

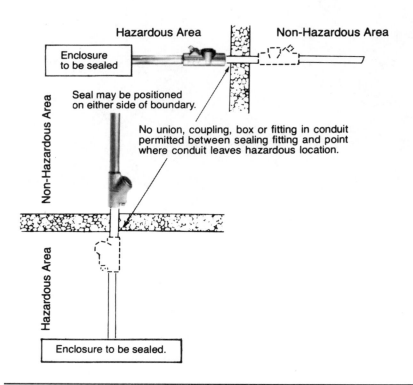

Hazardous Area

Non-Hazardous Area

Enclosure to be sealed

Non-Hazardous Area

Seal may be positioned on either side of boundary.

No union, coupling, box or fitting in conduit permitted between sealing fitting and point where conduit leaves hazardous location.

Hazardous Area

Enclosure to be sealed.

Figure 7–5 Seal fitting installation guidelines and examples of fittings. (Courtesy of Appleton Electric Company.)

Sealing Requirements

Approved seals are required in commercial garages just as they are for any Class I, Division 1 or 2 area. The seal fittings must be installed in accordance with the provisions of Section 501-5 for horizontal as well as vertical boundaries of the defined locations (see Figure 7–5).

Wiring above Class I Locations

The wiring above the Class I location must be installed in metal raceways, nonmetallic conduit, electrical nonmetallic tubing, flexible metal conduit, liquid-tight flexible metal conduit, and liquid-tight flexible nonmetallic conduit or shall be type MC, MI manufactured wiring systems, or PLTC cable in accordance with Article 725. TC cable, cellular metal floor raceways, or cellular concrete floor raceways are permitted to be used below the floor, but such raceways shall have no connections leading into or through the Class I location above the floor (see Figure 7–2).

Where a circuit is portable or supplied from pennants and includes a grounded conductor as provided in Article 200, receptacles, attachment plugs, connectors, and similar devices shall be of the polarized type, and the grounded conductor of the flexible cord must be connected to the screw shell of the lampholder or to the grounded terminal of any utilization equipment.

Equipment Located above Class I Locations

Where equipment is located above the Class I location, the equipment shall be at least 12 feet above the floor level. Where it is deemed arc-producing equipment that might produce arcs, sparks, or particles of hot metal such as cut-out switches, charging panels, generators, motors, or other similar equipment. Having make-and-break sliding contacts, it is required to be of the totally enclosed type or so constructed as to prevent the escape of sparks or hot metal particles (see Figure 7–2). This requirement does not include receptacles, lamps, or lampholders.

Fixed lighting as located over the lanes through which vehicles are commonly driven or otherwise exposed to physical damage shall be located not less than 12 feet above the floor unless totally enclosed or so constructed as to prevent the escape of arcs, sparks, or hot metal particles.

Battery chargers and their control equipment are not permitted to be located within the classified locations in Section 511-3.

Ground-Fault Circuit-Interrupter Protection

Section 511-10 requires that all 125-volt single-phase 15- and 20-ampere receptacles installed in areas where electrical diagnostic equipment, electrical hand tools, or portable lighting equipment are to be used have ground-fault circuit-interrupter protection for personnel.

Although this article covers only specific areas, it can provide good guidelines for similar areas in which other combustible engines are being serviced and repaired. Bear in mind that boat shops, lawnmower shops, and so on, although similar in nature, must comply with Article 501 in areas classified as such.

Grounding

In addition to the general requirements of Article 250, which covers all areas in a commercial garage, the classified areas shall be grounded in accordance with Section 501-16.

CHAPTER

8

Aircraft Hangars

Aircraft hangars are defined by *NEC®* Article 513 as locations used for storage or servicing of aircraft in which Class 1 flammable liquids, Class II combustible liquids, gasoline, jet fuels, or other volatile, flammable liquids or flammable gases are housed or stored (see Figure 8–1). It shall not include locations used exclusively for aircraft that have never contained such liquids or gases or that have been drained and properly purged. This includes areas such as assembly plants and rebuild plants for aircraft. Although it is becoming increasingly popular for residential developments to be built in which individual aircraft are stored adjacent to the residence, this article has not been revised to cover those areas, and therefore, where encountered, they must meet the provisions of *NEC®* Article 513. Applicable building codes might also have additional requirements related to these special residential developments with an airstrip and individual aircraft hangar on each resident's property.

Specific designated areas within each aircraft hangar are classified according to the degree of hazard that might exist generally. The authority having jurisdiction has the final determination regarding the classification of locations. However, the requirements of the *NEC®* are the minimal requirements and are generally applicable.

Class I, Division 1 Areas

Any pit or depression below the level of the hangar floor extending up to floor level.

Class I, Division 2 Areas

These areas include the entire area of the hangar, including any adjacent and communicating areas not suitably cut off from the hangar up to a level 18 inches (457 mm) above the floor and the area within 5 feet (1.52 m) horizontally from aircraft power plants (engines) and/or aircraft fuel tanks extending upward from the floor to a level 5 feet (1.52 m) above the upper surface of wings and/or engine enclosures (see Figure 8–2).

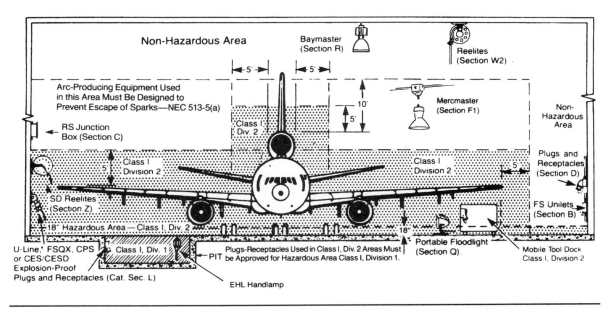

Figure 8–1 Recommended installation guidelines for aircraft hangars. (Courtesy of Appleton Electric Company.)

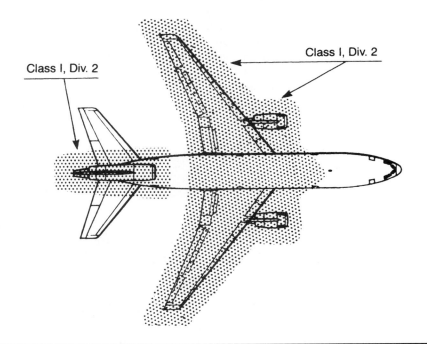

Figure 8–2 Top view of Class I, Division 2 areas. (Courtesy of Appleton Electric Company.)

Nonclassified Areas

Any areas that are suitably cut off from the classified portion of the hangar by walls or partitions and are ventilated so that flammable liquids or vapors are not likely to be present (i.e., stockrooms, electrical control rooms, and other similar locations) need not be classified.

Wiring and Equipment in Classified Locations

All wiring and equipment that is or might be installed and/or operated within hangar areas that are classified Class I, Divisions 1 or 2 shall comply with the applicable provisions of Article 501. Wiring installed in or under the hangar floor must comply with the requirements for Class I, Division 1. Adequate drainage is required where classified wiring is located in vaults, pits, or ducts. The wiring shall not be installed in enclosures or compartments with any service other than piped compressed air. Attachment plugs and receptacles in classified locations shall be of a type approved for the location in accordance with Article 501 (see Figure 8–3). Any jet-fueling station that is normally located near the aircraft, the hangars, and runways must comply with Articles 514 and 515.

Wiring and Equipment Not within Class I Locations

All fixed wiring in a hangar but not within a classified location shall be installed in metal raceways or be type MI, TC, or MC cable (see Figure 4–15 in Chapter 4 and Figure 7–4 in Chapter 7). Wiring suitably cut off from the hangar area 3 can be of the other suitable wiring methods in *NEC*® Chapter 3.

Where portable equipment, lamps, or pendants are installed, they must be of a flexible cord suitable for the type of service, and the cord must contain a separate green-equipment-grounding conductor.

For circuits supplying portable equipment or pendants, including a grounded conductor as provided in Article 200, the grounded conductor is to be connected to the screw shell of lampholders and to the grounded terminal of all utilization equipment supplied. All receptacles, attachment plugs, connectors, and similar devices must be the grounding type. An approved means must be provided for maintaining continuity of the grounding con-

Figure 8–3 Typical jet fuel bulk fuel station. (Courtesy of Appleton Electric Company.)

ductor between the fixed wiring system and the non-current-carrying metal portions of pendant fixtures, portable lamps, and portable utilization equipment.

Equipment that is less than 10 feet above wings and/or engine enclosures of aircraft but is not within the classified area as described in the *NEC*® that might produce arcs, sparks, or particles of hot metal, such as lampholders for fixed lighting, switches, receptacles, or other equipment having make-and-break or sliding contacts, must be of the totally enclosed type or designed to prevent escape of sparks or hot metal particles. General purpose types of equipment can be used in those areas described in Section 513-3(d); however, some restrictions apply to the specific type of lampholders and portable equipment that can be used.

Stanchions, Rostrums, and Docks

All electric wiring, outlets, and equipment (including lamps) on or attached to stanchions, rostrums, or docks that are or are likely to be located in a Class I location 3 must comply with the requirements for Class I, Division 2 locations. Where they are not and are not likely to be located in a Class I location, wiring and equipment located more than 18 inches above the floor need only comply with Sections 513-5 and 513-6. Wiring and equipment not more than 18 inches above the floor in any position shall comply with the requirements of Section 513-5(a), and all receptacles and attachment plugs must be of the locking type. Mobile stanchions not intended to be used within the classified areas are required to have at least one permanently affixed warning sign to read: "**Warning:** Keep 5 feet clear of aircraft engines and fuel tank areas."

Sealing

The rules for sealing in aircraft hangars are defined by class and division (see Figure 7–5 in Chapter 7). Seals must be installed in accordance with the provisions of the *NEC*®. These requirements apply to horizontal as well as to vertical boundaries of the defined Class I locations. Raceways embedded in a concrete floor or buried beneath a floor are considered to be within the Class I location.

Additional Specific Requirements

Article 513 requires that all aircraft electrical systems be de-energized when the aircraft is stored in a hangar. This section recommends that systems be deenergized during maintenance operations whenever possible.

Aircraft batteries shall not be charged where installed in an aircraft located inside or partially inside a hangar. Battery chargers and their control equipment are not permitted to be located or operated within any of the Class I locations.

Tables, racks, trays, and wiring are not permitted within a Class I location and, in addition, must comply with Article 480 for storage batteries.

It is recommended that charging equipment be located in a separate building or in a nonclassified area. All mobile chargers are required to have at least one permanently affixed warning sign to read: "**Warning:** Keep 5 feet clear of aircraft engines and fuel tank areas."

Aircraft energizers (external power sources) shall be so designed that all electric equipment and fixed wiring will be at least 18 inches above floor level and must not be operated in a Class I location. Mobile energizers must have at least one permanently affixed warning sign to read: "**Warning:** Keep 5 feet clear of aircraft engines and fuel tank areas."

Mobile servicing equipment (e.g., vacuum cleaners, air compressors, and air movers) having electric wiring and equipment not suitable for Class I, Division 2 locations must be so designed that all fixed wiring and equipment will be at least 18 inches above the floor. This equipment is not permitted to be operated within the Class I location and must have at least one permanently affixed warning sign to read: "**Warning:** Keep 5 feet clear of aircraft engines and fuel tank areas."

Equipment that is not identified as suitable for Class I, Division 2 locations is not permitted to be operated in locations where, during maintenance operations, flammable liquids or vapors are likely to be released.

Grounding

All metal raceways, metal-jacketed cables, and all non-current-carrying metal portions of fixed or portable equipment, regardless of voltage, shall be grounded as provided in Article 250.

CHAPTER

9

Gasoline Dispensing and Service Stations

Gasoline dispensing and service stations are defined in *NEC®* Article 514 as a location where gasoline or other volatile, flammable liquids or liquefied flammable gases are transferred to fuel tanks, including auxiliary fuel task of self-propelled vehicles or approved containers. These are areas where flammable liquids having a flash point below 38°C (100°F), such as gasoline or liquid petroleum gas, are dispensed. These are generally areas within a facility and not the entire facility. For example, a truck stop or combination-use facility might contain several type areas other than a gasoline dispensing and service station, such as a place of assembly as covered in Article 518 (e.g., a restaurant). It might also contain an area that is classified as a commercial garage as covered in Article 511 and, if suitably separated, a diesel fuel-dispensing area that is not required to comply with the provisions of this article. Resorts commonly have more than one area coming under the provisions of this article, such as a gasoline-dispensing island for retail sales and one for their own equipment, a propane tank fill area, and a marine area for fueling boats. Other areas, such as lubritoriums, service rooms, repair rooms, offices, sales rooms, compressor rooms, and similar locations, shall comply with *NEC®* Articles 510 and 511 with respect to the electrical wiring and equipment. Where the authority having jurisdiction can satisfactorily determine that flammable liquids having a flash point below 38°C (100°F), such as gasoline, will not be handled, such locations need not be classified. Additional information regarding the safeguards for gasoline dispensing and service stations can be obtained by seeing the Automotive and Marine Service Station Code NFPA 30. For additional information pertaining to LP gas systems other than residential or commercial, see NFPA 58. For storage and handling of liquefied petroleum gases, see NFPA 59. For gasoline-dispensing stations and marine applications, such as boatyards, see *NEC®* Article 555. It is important that each area be classified separately to ensure a safe installation that will remain safe throughout the life of the facility with proper maintenance.

Class I Locations

The areas within the scope of Article 514 that are to be classified Class I are clearly defined in Table 514-2, and the boundaries for each area to be classified does not extend beyond the unpierced wall, the roof, or other solid partitions of that area. In open-air areas, the classified area boundaries are measured in feet or inches from the hazardous substance source, such as the fill opening, dispenser, or vent discharge. The likelihood of the presence of the liquid or vapors dictates the division. The group for these materials is Group D. All wiring and electrical equipment covered within these classified Class I locations are required to comply with the applicable provisions of Article 501 except where other wiring methods are specifically allowed.

The space within a dispenser up to a height of 4 feet and the space within 18 inches horizontally of a dispenser and extending to a height 4 feet above its base must be classified as a Class I, Division 1 location (see Figure 9–1). Space classification inside the dispenser enclosure is covered in *Power Operated Dispensing Devices for Petroleum Products, ANSI/UL87.*

Any wiring or equipment that is installed beneath any part of a Class I, Division 1 or 2 location is classified as being within the Class I, Division 1 location to the point of emergence from below grade. Where the dispensing unit or its hose or nozzle valve has not been suspended from overhead, the space within the dispenser enclosure and the area within 18 inches in all directions from the enclosure that is not suitably cut off by a ceiling or wall is to be classified as a Class I, Division 1 area.

All equipment integral with the overhead dispensing hose and nozzle must be suitable for Class I, Division 1 hazardous locations. A spherical volume within a 3-foot radius of the point of discharge of any tank vent pipe is classified as a Class I, Division 1 location.

Lubritoriums, service rooms, and repair rooms, as well as other areas not suitably cut off or elevated at least 18 inches above these areas, are classified and considered as commercial garages. These areas generally contain areas that are classified as Class I, Division 1, and except for pits and depressions in floors, many of these areas, when properly designed, can be reduced to Division 2 or nonclassified.

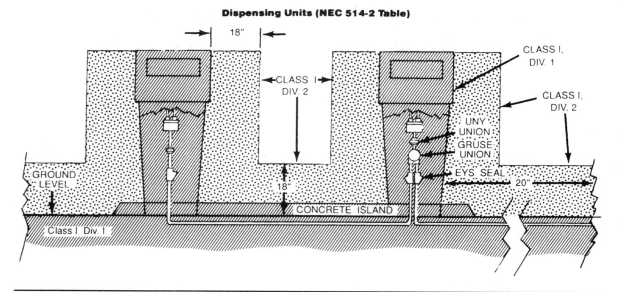

Figure 9–1   Example of a typical gasoline island area. (Courtesy of Appleton Electric Company.)

Class I, Division 2

The area within 2 feet in all directions of the Class I, Division 1 area and extending down to grade level is classified as a Class I, Division 2 area. The horizontal area 18 inches above grade and extending 20 feet measured from those points vertically below the outer edges of the overhead dispensing enclosure is also classified as a Class I, Division 2 area (see Figures 9–2 and 9–3).

The area within 3 feet measured in any direction from the dispensing point of a hand-operated unit dispensing a Class I flammable liquid, such as white gas, is considered a

Table 514-2 Class I Locations—Service Stations

Location	Class I, Group D Division	Extent of Classified Location
Underground Tank		
Fill Opening	1	Any pit, box, or space below grade level, any part of which is within the Division 1 or 2 classified location.
	2	Up to 18 in. above grade level within a horizontal radius of 10 ft from a loose fill connection and within a horizontal radius of 5 ft from a tight fill connection.
Vent—Discharging	1	Within 3 ft of open end of vent, extending in all directions.
Upward	2	Space between 3 ft and 5 ft of open end of vent, extending in all directions.
Dispensing Device[1,4]		
(except overhead type)[2]		
Pits	1	Any pit, box, or space below grade level, any part of which is within the Division 1 or 2 classified loation.
Dispenser		FPN: Space classification inside the dispenser enclosure is covered in *Power Operated Dispensing Devices for Petroleum Products,* ANSI/UL 87-1995.
	2	Within 18 in. horizontally in all directions extending to grade from the dispenser enclosure or that portion of the dispenser enclosure containing liquid handling components. FPN: Space classification inside the dispenser enclosure is covered in *Power Operated Dispensing Devices for Petroleum Products,* ANSI/UL 87-1995.
Outdoor	2	Up to 18 in. above grade level within 20 ft horizontally of any edge of enclosure.
Indoor		
with Mechanical Ventilation	2	Up to 18 in. above grade or floor level within 20 ft horizontally of any edge of enclosure.
with Gravity Ventilation	2	Up to 18 in. above grade or floor level within 25 ft horizontally of any edge of enclosure.

(continues)

Figure 9–2   Table 514-2. (Reprinted with permission from NFPA 70-199, *National Electrical Code®,* Copyright © 1998, National Fire Protection Association, Quincy, MA 02269.)

Location	Class I, Group D Division	Extent of Classified Location
Dispensing Device[4]		
Overhead Type[2]	1	The space within the dispenser enclosure, and all electrical equipment integral with the dispensing hose or nozzle.
	2	A space extending 18 in. horizontally in all directions beyond the enclosure and extending to grade.
	2	Up to 18 in. above grade level within 20 ft horizontally measured from a point vertically below the edge of any dispenser enclosure.
Remote Pump—Outdoor	1	Any pit, box, or space below grade level if any part is within a horizontal distance of 10 ft from any edge of pump.
	2	Within 3 ft of any edge of pump, extending in all directions. Also up to 18 in. above grade level within 10 ft horizontally from any edge of pump.
Remote Pump—Indoor	1	Entry space within any pit.
	2	Within 5 ft of any edge of pump, extending in all directions. Also up to 3 ft above grade level within 25 ft horizontally from any edge of pump.
Lubrication or Service Room— With Dispensing	1	Any pit within any unventilated space.
	2	Any pit with ventilation.
	2	Space up to 18 in. above floor or grade level and 3 ft horizontally from a lubrication pit.
Dispenser for Class I Liquids	2	Within 3 ft of any fill or dispensing point, extending in all directions.
Lubrication or Service Room— Without Dispensing	2	Entire area within any pit used for lubrication or similar services where Class I liquids may be released.
	2	Are up to 18 in. avove any such pit, and extending a distance of 3 ft horizontally from any edge of the pit.
	2	Entire unventilated area within any pit, below grade area, or subfloor area.
	2	Area up to 18 in. above any such unventilated pit, below grade work area, or subfloor work area and extending a distance of 3 ft horizontally from the edge of any such pit, below grade work area, or subfloor area.
	Nonclassified	Any pit, below grade work area, or subfloor work area that is provided with exhaust ventilation at a rate of not less than 1 cfm/ft² (0.3 m³/minute/m²) of floor area at all times that the building is occupied or when vehicles are parked in or over this area and where exhaust air is taken from a point within 12 in. (0.3 m) of the floor of the pit, below grade work area, or subfloor work area.

Location	Class I, Group D Division	Extent of Classified Location
Special Enclosure Inside Building[3]	1	Entire enclosure.
Sales, Storage, and Rest Rooms	Nonclassified	If there is any opening to these rooms within the extent of a Division 1 location, the entire room shall be classified as Division 1.
Vapor Processing Systems Pits	1	Any pit, box, or space below grade level, any part of which is within a Division 1 or 2 classified location or that houses any equipment used to transport or process vapors.
Vapor Processing Equipment Located Within Protective Enclosures FPN: See *Automotive and Marine Service Station Code,* NFPA 30A-1996, Section 4-5.7	2	Within any protective enclosure housing vapor processing equipment.
Vapor Processing Equipment Not Within Protective Enclosures (excluding piping and combustion devices)	2	The space within 18 in. in all directions of equipment containing flammable vapor or liquid extending to grade level. Up to 18 in. above grade level within 10 ft horizontally of the vapor processing equipment.
Equipment Enclosures	1	Any space within the enclosure where vapor or liquid is present under normal operating conditions.
Vacuum-Assist Blowers	2	The space within 18 in. in all directions extending to grade level. Up to 18 in. above grade level within 10 ft horizontally.

Note: For SI units, 1 in. = 2.5 cm; 1 ft = 0.3048 m.

[1]Refer to Figure 514-2 for an illustration of classified location around dispensing devices.

[2]Ceiling mounted hose reel.

[3]FPN: See *Automotive and Marine Service Station Code,* NFPA 30A-1996, Section 2-2.

[4]FPN: Area classification inside the dispenser enclosure is covered in *Power-Operated Dispensing Devices for Petroleum Products,* ANSI/UL 87-1995.

Figure 9–2 *Continued.*

Location

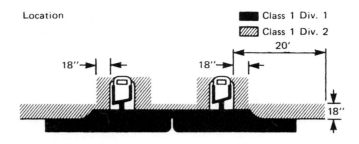

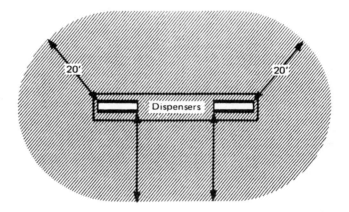

Figure 1. Classified Locations Adjacent to Dispensers as Detailed in Table 514-2.

Exception No. 1: Type MI cable shall be permitted where it is installed in accordance with Article 330.

Exception No. 2: Rigid nonmetallic conduit complying with Article 347 shall be permitted where buried under not less than 2 feet (610 mm) of earth. Where rigid nonmetallic conduit is used, threaded rigid metal conduit or threaded steel intermediate metal conduit shall be used for the last 2 feet (610 mm) of the underground run to emergence or to the point of connection to the aboveground raceway; an equipment grounding conductor shall be included to provide electrical continuity of the raceway system and for grounding of noncurrent-carrying metal parts.

Figure 9–3 Figure 1 detailed to Table 514-2. (Reprinted with permission from NFPA 70-1999, *National Electrical Code®*, Copyright © 1998, National Fire Protection Association, Quincy, MA 02269.)

Class I, Division 2 location. An outside location in buildings not suitably cut off or any area not classified as a Class I, Division 1 location that is within 20 feet horizontally of the interior enclosure of the dispensing pump or within 10 feet horizontally from a tank filled by and less than 18 inches above grade is to be classified as a Class I, Division 2 location (see Figure 9–1). That part of the spherical volume lying between a 3-foot radius and a 5-foot radius of the point of discharge of any tank vent pipe is classified as a Class I, Division 2 location. If the tank vent does not discharge upward, the area extending to grade level is classified as a Class I, Division 2 location.

Areas Not Classified

The space beyond the 5-foot radius from the tank vents that discharge upward and spaces beyond unpierced walls and areas below grade that lie beneath tank vents are not classified as hazardous.

Seals

A seal is required as the first fitting after a conduit emerges from the earth or concrete and enters a dispenser or any cavity or enclosure in connection with it. A seal is required in any conduit that extends upward beyond the vertical boundary of a Class I, Division 2 area, such as a conduit supplying a lighting standard within a hazardous area (see Figure 7–5 in Chapter 7). The required seal must be placed so that no coupling or fitting of any kind is between the seal and the boundary line dividing the hazardous and nonhazardous areas. A seal is required at or beyond the horizontal boundary at the point where the conduit emerges from below grade. The seal must be the first fitting where the conduit emerges into a nonhazardous area. If the conduit does not extend beyond the 20-foot horizontal boundary but emerges and extends upward while still within the hazardous area, the seal must be so placed that no coupling or fitting is between it and the vertical boundary. Circuits leading to or through a gasoline dispenser enclosure (see Figure 9–1) must be provided with switches or circuit breakers that simultaneously disconnect all circuit conductors, including the neutral, from the source of supply. Section 514-5 states that single pole breakers utilizing handle ties are not permitted for this application.

Warning: Unless specifically marked, seals are limited to a 25% wire fill. Generally, the code will permit an oversized seal to achieve the 40% wire fill allowed for the conduit. Some manufacturers are now producing a seal that will permit a 40% wire fill.

Circuit Disconnects

A disconnecting means is required for each circuit leading to or through all dispensing equipment. The switch or other acceptable means must disconnect simultaneously all conductors, including the grounded conductor. Specific rules for stations that have attendants on duty and those that are unattended as to the type and location of required emergency controls can be found in NFPA 30A. Extracts from this standard can be found in Section 514-5(b) and (c).

CHAPTER

10

Bulk Storage Plants

Definition

A bulk storage plant is defined as a portion of the property where bulk fuels are received and distributed. While at that location, they might also receive additives or be blended to achieve the desired mixture requested. These fuels can be received and distributed in any number of ways, such as pipelines, railcar tanks, tank ships, or tank trucks. For further information, check the latest Flammable and Combustible Liquids Code NFPA 30-1996 (ANSI) (see Figures 10–1 and 10–2).

The degree of hazard or flammability of liquids is commonly measured in terms of the flash point. The flash point is the minimum temperature at which the liquid gives off vapors sufficient to form an ignitable mixture with air near the surface of the liquid when tested under established procedures. Flammable liquids having a flash point less than 100°F require the greatest protection against accidental ignition. Gasoline, for example, has a flash point of about 45°F and is a very hazardous material under all climatic conditions. Diesel fuel has a flash point of 100°F. The flash points of jet fuels range from 10°F to 145°F. See Fire Hazard Properties in Flammable Liquids, Gases, and Volatile Solids.

Figure 10–1 Photo of bulk storage tank. (Courtesy of Appleton Electric Company.)

Class I Locations

Article 514 applies in areas or rooms where Class I liquids are handled and stored and where gasoline is dispensed in conjunction with the bulk storage operations. The boundaries of these areas or rooms are defined by the floors, walls, roofs, or other solid partitions without communicating openings. *NEC*® Table 515-2 can be used to classify bulk storage facilities, provided that these areas meet all the requirements of NFPA 30 and Article 501. The authority having jurisdiction can specify the extent of these spaces. *NEC*® Article 515 does not cover necessary physical and corrosion protection needs.

It is important that designers understand that the *NEC*® Chapter 5 articles deal with the special considerations necessary to achieve a safe system in hazardous locations. However, other considerations might be necessary to achieve an installation that will last many years in the applicable environment. As an example, in many bulk storage tank farms, corrosion can be an important element to consider. Petroleum products and corrosive chemicals can greatly reduce the life of the wiring method if not considered. Physical protection can also be an important factor in these areas. When both are present, the wiring method options are greatly limited. Coated galvanized rigid conduit might be advisable because this wiring method provides both good physical protection and excellent corrosion protection.

Adequately ventilated indoor areas containing pumps, bleeders, withdrawal fittings, meters, and similar equipment that are connected to pipelines handling flammable liquids under pressure are considered as Class I, Division 2 locations where such spaces are within a 5-foot radius in any direction of an exterior surface of such equipment. A Class I, Division 2 location also extends 25 feet horizontally from the surface of such equipment and upward to a height of 3 feet above grade (see Figure 10–3). Where such spaces are inadequately vented, they are considered a Class I, Division 1 location. Other provisions affecting ventilating requirements are discussed in the flammable liquids code, NFPA 30. Outdoor areas containing the same equipment itemized in the preceding paragraph are to be considered as Class I, Division 2 locations within a 3-foot radius in all directions from the exterior surfaces of such equipment. The Class I, Division 2 location also extends 10 feet horizontally from any surfaces of such equipment and extends upward to a height of

Figure 10–2 Photo of a bulk storage and loading facility. (Courtesy of Crouse-Hinds, Division of Cooper Industries.)

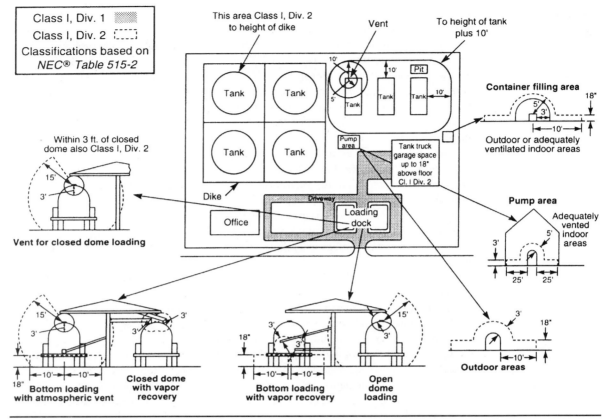

Figure 10–3 Bulk storage installation guidelines. (Courtesy of Crouse-Hinds, Division of Cooper Industries.)

18 inches above grade. In *NEC*® Table 515-2 and Figure 10–3, such spaces are considered to be Class I, Division 1 locations within 3 feet in all directions of the vent pipe or fill pipe opening. Class I, Division 2 locations are within the space extending in all directions between a 3-foot and 5-foot radius from the vent pipe or fill pipe opening and horizontally within the radius of 10 feet from the vent or fill pipe opening to a height of 18 inches above grade. In all indoor areas where volatile, flammable liquids are transferred to individual containers or where fire-enforced ventilation is not reliably maintained, the entire area involved must be classified as a Class I, Division 1 location (*NEC*® Table 515-2[b]).

More limited hazardous locations are also defined for outdoor areas where volatile, flammable liquids are transferred to individual containers as well as for indoor areas that have reliably maintained fire-enforced ventilation. In outside locations, the space extending 3 feet in all directions from a vehicle tank when loading through an open dome or from a vent when loading through a closed dome having atmospheric venting is classified as a Class I, Division 1 location. Between a 3-foot and 5-foot radius, it is a Class I, Division 2 location. Hazardous areas surrounding loading facilities and tank facilities are defined in *NEC*® Table 515-2. The space within 3 feet in all directions of a fixed connection used in bottom loading or unloading is a Class I, Division 1 location. In loading through a closed dome having atmospheric venting or in loading through a closed dome having a vapor-recovery system, it is considered a Class I, Division 2 location. Where bottom loading is used, the Class I, Division 2 location covers a 10-foot radius from the point of connection to a height of 18 inches above grade (see Figure 10–3). In above-ground tanks, the spaces between the shell and above the roof of floating roof tanks are Class I, Division 1 locations. The area within 10 feet of where the shell sides end and the roof of the above-ground tank that does not have a floating roof is classified as a Class I, Division 2 location. Where dikes are provided, the area within the dike is also a Class I, Division 2 location. Any point within a 5-foot radius of a vent opening for an above-ground tank is a Class I, Division 1 location. The space between a 5-foot and 10-foot radius from the vent of an above-ground tank is a Class I, Division 2 location. Hazardous locations near fill pipes and vent pipes for underground tanks are defined in *NEC*® Table 514-2 (see Figure 10–3). Any pit or depression that lies wholly or partly within either a Class I, Division 1 or Class I, Division 2 location must be classified entirely as a Class I, Division 1 location unless provided with forced-draft ventilation. Where such forced-draft ventilation is provided, all such spaces are classified as a Class I, Division 2 location. Any pit or depression that contains gasoline piping, valves, or fittings is deemed a Class I, Division 2 location when located in areas that would not otherwise be classified as a Class I area, such as where a gasoline pipeline or the pit outside the bulk storage plant contains meters or valves. Storage and repair garages for tank trucks are classified as commercial garages; that is, the Class I, Division 2 area extends up to a height of 18 inches above grade unless the enforcement authority rules that the hazardous area must extend higher than 18 inches. Wiring within hazardous areas is required to comply with the *NEC*® requirements for Class I, Division 1 or Class I, Division 2.

Warning: Great care must be exercised around waterways when loading and unloading flammable fluids. Careful consideration must be given to the hazards involved, plus corrosion and environmental concerns (see Figure 10–4).

Class I, Divisions 1 and 2 Wiring and Equipment

All defined classified locations shall be in compliance with the provisions of *NEC*® Table 515-2.

Above the hazardous areas, fixed equipment installed in bulk storage plants is required to be of the totally enclosed type if capable of producing arcs or sparks, unless provided with screens or guards to prevent the escape of the hot particles produced. All portable equipment, such as portable lamps and their cords, is required to comply with Article 501. Fixed wiring shall be used in metal raceways, Schedule 80 rigid nonmetallic conduit, and cable types MI, MC, TC, and MV.

Underground wiring is required to be installed with threaded rigid metal conduit or threaded steel intermediate conduit or where buried not less than 2 feet below the earth and is permitted to be in rigid nonmetallic conduit or an approved cable. Where rigid nonmetallic conduit is used, threaded rigid metal conduit or threaded intermediate metal conduit shall be used the last 2 feet of the conduit run to emergence to the point of connection to the above-ground raceway. Where cable is used, it must be enclosed in threaded rigid metal conduit or threaded steel intermediate metal conduit from the point of the lowest buried cable level to the point of connection above the raceway.

Conductor insulation in all underground wiring must comply with Section 501-13. An equipment-grounding conductor to ground all non-current-carrying metal parts is required in all nonmetallic wiring methods.

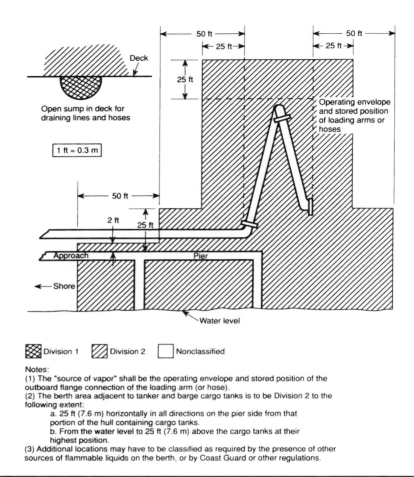

Division 1   Division 2   Nonclassified

Notes:
(1) The "source of vapor" shall be the operating envelope and stored position of the outboard flange connection of the loading arm (or hose).
(2) The berth area adjacent to tanker and barge cargo tanks is to be Division 2 to the following extent:
     a. 25 ft (7.6 m) horizontally in all directions on the pier side from that portion of the hull containing cargo tanks.
     b. From the water level to 25 ft (7.6 m) above the cargo tanks at their highest position.
(3) Additional locations may have to be classified as required by the presence of other sources of flammable liquids on the berth, or by Coast Guard or other regulations.

Figure 10–4 Marine wharf or loading and unloading dock area. (Reprinted with permission from NFPA 70-1999, *National Electric Code®,* Copyright © 1998, National Fire Protection Association, Quincy, MA 02269.)

Sealing

Sealing fittings are required in electrical systems in accordance with the *NEC®* provisions for Class I areas—that is, in all raceways entering or leaving a hazardous location, all raceways entering or leaving an arcing device, and all raceways 2 inches in size or larger entering or leaving a junction box and containing terminals, taps, or splices.

Nonhazardous Locations

Office areas, boiler rooms, and similar locations beyond the limits of the hazardous area are not required to come within the special wiring rules of this article.

Grounding

Article 250 applies to all areas within the scope of Article 515. Class I areas must also comply with the special grounding requirements in Article 501.

     Static electricity can build up to dangerous levels from truck tires and should be considered. Manufacturers can provide a safe means for static discharge.

CHAPTER

11

Spray Application, Dipping, and Coating Processes

Defined Location

These locations include both Class I and Class II areas, depending on the application. The application of flammable liquids, combustible liquids, and combustible powders by spray operations and the application of flammable liquids or combustible liquids at temperatures above their flash points by dipping, coating, and other means regularly or frequently are covered in *NEC®* Article 516. The classification is based on quantities of dangerous vapors, combustible dusts, mists, residues, or deposits. Additional information regarding the safeguards for these processes is to be found in NFPA 33-1995, NFPA 34-1996, and NFPA 91-1995.

These finishes can be applied by means of dipping, brushing and spraying, pneumatic spraying, fixed electrostatic spraying, electrostatic hand spraying, electrostatic fluidized beds, and powder sprying with electrostatic power spraying guns.

Two distinct hazards are present in paint spraying operations: explosive mixtures in the air from the spray and its vapor and the combustible residue that can accumulate inside the booth (see Figures 11–1 and 11–2).

Class I or Class II, Division 1 Locations

Section 516-2(a) states that the interior of spray booths and their exhaust ducts and all spaces within 20 feet horizontally in any direction from the spraying operation are classified as a Class I, Division 1 location. The areas where spraying more extensive than touch-up spraying operations are performed, and all space within 20 feet horizontally in any direction of dip tanks and their drainboards are classified as a Class I, Division 1 location.

These particular facilities can contain both Class I and Class II requirements, depending on the material that is being applied:

- Interiors of spray booths and rooms
- The interior of exhaust ducts
- Areas in the direct path of spray operations
- Areas for dipping and coating operations
- All space within 5-foot radial distance from the vapor sources extending from these surfaces to the floor

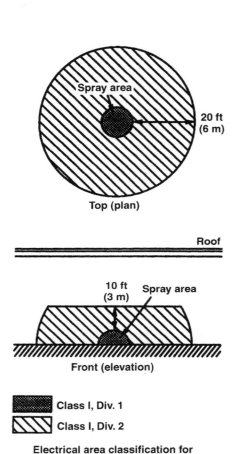

Top (plan)

Roof

10 ft
(3 m)   Spray area

Front (elevation)

Class I, Div. 1

Class I, Div. 2

Electrical area classification for
open spray areas.

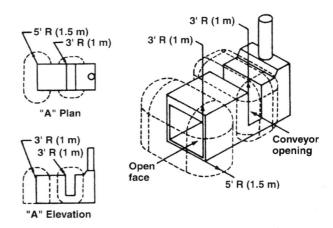

"A" Plan

"A" Elevation

Electrical area classification for open-faced or
open-front spray booth or spray room where exhaust ventilation
is interlocked with spray application equipment.

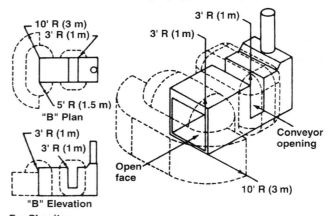

"B" Plan

"B" Elevation

For SI units:
one inch = 25.4 millimeters;
one foot = 0.3048 meter.

Electrical area classification for open-faced or
open-front spray booth or spray room where exhaust ventilation
is not interlocked with spray application equipment.

Figure 11–1 Class I or Class II location adjacent to an unenclosed spray operation. (Reprinted with permission from NFPA 70-1999, *National Electrical Code*®, Copyright © 1998, National Fire Protection Association, Quincy, MA 02269.)

Figure 11–2 Class I or Class II, Division 2 location adjacent to a closed-top, open-face, or open-front spray booth or room. (Reprinted with permission from NFPA 70-1999, *National Electrical Code*®, Copyright © 1998, National Fire Protection Association, Quincy, MA 02269.)

The vapor source shall be a liquid exposed in the process and the drainboard and any dipped or coated object from which it is possible to measure the vapor for concentrations exceeding 25% of the lower flammable limit at a distance of 1 foot in any direction, pits within 25 feet horizontally of the vapor source (if the pit extends beyond 25 feet, a vapor stop must be provided or the entire pit must be classified), and the interior of any enclosed coating or dipping process.

Class I or Class II, Division 2 Locations

These locations include the following spaces, again depending on the material being applied, as to whether it would be a Class I or Class II location: for open spraying, all space outside but within 20 feet horizontally and 10 feet vertically of the Division 1 location as defined and not separated by partitions (see Figure 11–1). For spraying operations conducted within an enclosed spray booth or room, the space shown in Figure 11–2 and the space within 3 feet in all directions from the openings other than the open face or front shall be considered a Class I and II, Division 2. Class I and II, Division 2 locations shown in Figure 11–2 shall extend from the spray booth or room as follows: If the ventilation system is interlocked with the spraying equipment so as to make the spraying equipment inoperative when the ventilation system is not in operation, the space shall extend 5 feet from the open face or front of the booth or room, as otherwise shown in Figure 11–2. When the ventilation system is not interlocked with the spraying equipment, the space shall extend 10 feet from the open face or front of the spray booth (see Figure 11–2). For spraying operations conducted within an open-top spray booth, the space 3 feet vertically above the booth and within 3 feet of other booth openings shall be considered a Division 2 location. For spraying operations confined to an enclosed booth or room, the space within 3 feet in all directions from any openings is considered a Division 2 location, as shown in Figure 11–2. For dip tanks and drainboards, the space 3 feet above the floor and extending 20 feet horizontally in all directions from the Division 1 location shall be classified a Class I or II, Division 2 location, except that the space would not be required to be hazardous where the vapor source area is 5 square feet or less and where the contents of the open tank, trough, or container do not exceed 5 gallons. For the classification of open processes, see Figure 11–5. For the classification of open-top containers, see Figure 11–6. In addition, the vapor concentration during operation shutdown periods must not exceed 25% of the lower flammable limit outside the Class I location as specified in the *NEC*®. In an enclosed coating and dipping operation (see Figures 11–4 and 11–7), the space adjacent to these operations is not considered to be classified except within 3 feet in all directions of any opening or enclosure that must be classified as a Division 2 location (see Figure 11–3).

Nonhazardous Locations

Adjacent areas that are cut off by tight partitions, without communicating openings, and within which hazardous vapors or combustible powders are not likely to be released might not be required to be classified. Areas utilizing drying, curing, or fusion apparatus provided with positive mechanical ventilation adequate to prevent accumulation of flammable concentrations of vapors and provided with effective interlocks to de-energize all electrical equipment other than the equipment approved for Class I locations might be classified as nonhazardous if the authority having jurisdiction so judges (see Figure 11–3). Information pertaining to drying ovens or furnaces can be found in NFPA 86-1990.

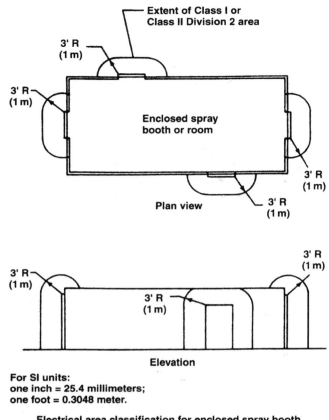

Figure 11-3 Class I or Class II, Division 2 location adjacent to openings in an enclosed spray booth or room. (Reprinted with permission from NFPA 70-1999, *National Electrical Code®*, Copyright © 1998, National Fire Protection Association, Quincy, MA 02269.)

Wiring and Equipment in Class I Locations

All electric wiring within the classified location as defined in Article 516 where *only vapors* are contained (*no residue*) must comply with the applicable provisions of *NEC®* Article 501. There shall be no electrical equipment located where dangerous deposits of combustible or flammable vapors, mists, residue, dusts, or deposits might be present unless specifically "listed" for the location and purpose. Only electrical wiring in rigid metal conduit or intermediate metal conduit with metal boxes is permitted. The boxes are not permitted to contain terminations, splices, or taps.

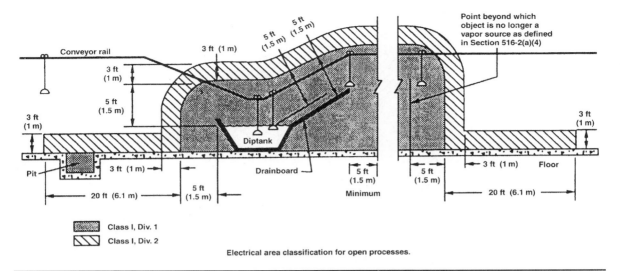

Electrical area classification for open processes.

Figure 11–4 The extent of a Class I, Division 1 and Class I, Division 2 locations for an open-dipping process. (Reprinted with permission from NFPA 70-1999, *National Electrical Code*®, Copyright © 1998, National Fire Protection Association, Quincy, MA 02269.)

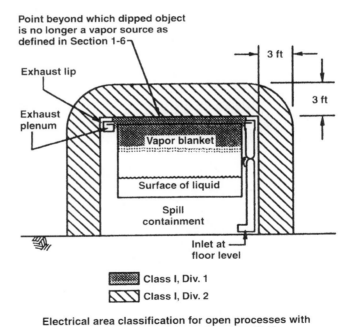

Electrical area classification for open processes with peripheral vapor containment and ventilation.

Figure 11–5 Area classification for open-process operations. (Reprinted with permission from NFPA 70-1999, *National Electrical Code*®, Copyright © 1998, National Fire Protection Association, Quincy, MA 02269.)

NFPA 33 — A95 ROP

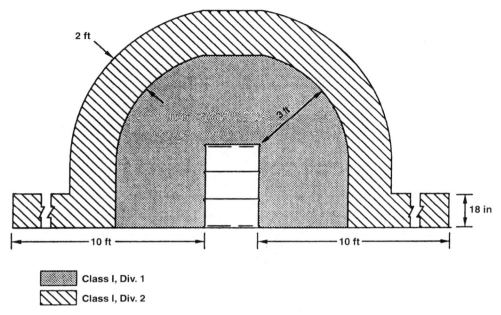

Electrical area classification around an open-top container.

Figure 11–6 Area classification for open-top containers. (Reprinted with permission from NFPA 70-1999, *National Electrical Code®,* Copyright © 1998, National Fire Protection Association, Quincy, MA 02269.)

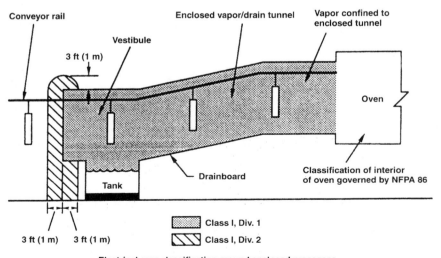

Electrical area classification around enclosed processes.

Figure 11–7 Area classification for enclosed processing operations. (Reprinted with permission from NFPA 70-1999, *National Electrical Code®,* Copyright © 1998, National Fire Protection Association, Quincy, MA 02269.)

Illumination

Specific requirements for lighting in these areas can be found in Article 516 and NFPA 33-1995. Paint spray booth manufacturers have struggled to meet the needs while complying with the requirements of this *NEC®* section. New and innovative methods have been employed to provide safety while meeting the needs of the customer. It has generally been the accepted rule that general purpose fixtures could be used if they were installed away from openings to the booth and nonbreakable clear panels isolated the fixture from the booth. Additionally, no access to service the fixture was required from inside the booth, and there was assurance that there would be no heat buildup on the inside face of the panels. However, this required that the booth have workable clearances around the outside of the booth, and that was not always obtainable. NFPA 33-1995 now provides reduced clearances and clear directions on these installation guidelines. The authority having jurisdiction should always be consulted before building or installing a booth that varies from the general requirements (see Figures 11–8 and 11–9).

Portable Equipment

Generally, portable equipment is not permitted to be used in a spray area during spraying operations:

- Section 516-3(d): Portable electric lamps or other utilization equipment shall not be used in a spray area during spray operations.
- Exception No. 1: Where portable electric lamps are required for operations in spaces not readily illuminated by fixed lighting within the spraying area, they shall be of the type approved for Class I, Division 1 locations where readily ignitable residues might be present.
- Exception No. 2: Where portable electric drying apparatus is used in automobile refinishing spray booths and the following requirements are met: (1) the apparatus and its electrical connections are not located within the spray enclosure during spray operations; (2) electrical equipment within 18 inches (45.7 cm) of the floor is

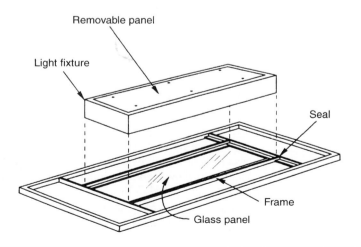

Figure 11–8   Example of a light fixture, frame, and seal installation in a paint spray box.

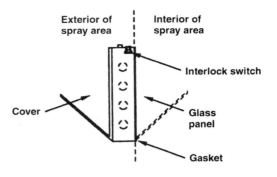

Exterior of spray area | Interior of spray area

Interlock switch

Cover

Glass panel

Gasket

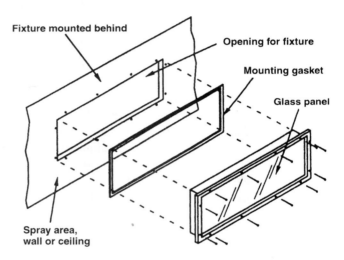

Fixture mounted behind

Opening for fixture

Mounting gasket

Glass panel

Spray area, wall or ceiling

Light fixture that is integral part of spray area and serviced from outside of spray area.

Figure 11–9 Sketch to show an acceptable method for installing illumination. (Reprinted with permission from NFPA 70-1999, *National Electrical Code®*, Copyright © 1998, National Fire Protection Association, Quincy, MA 02269.)

approved for Class I, Division 2 locations; (3) all metallic parts of the drying apparatus are electrically bonded and grounded; and (4) interlocks are provided to prevent the operation of spray equipment while drying apparatus is within the spray enclosure to allow for a 3-minute purge of the enclosure before energizing the drying apparatus and to shut off drying apparatus on failure of ventilation system.

- (e) Electrostatic Equipment. Electrostatic spraying or detearing equipment shall be installed and used only as provided in Section 516-4.

- (FPN): For further information, see Spray Application Using Flammable and Combustible Materials, NFPA 33-1995 (ANSI).

Special Equipment

The requirements for installing fixed electrostatic equipment, electrostatic hand spraying equipment, and the wiring of equipment for Class I or II locations and the areas above the Class I and Class II locations can be found in Sections 516-4(a) through (j), 516-5(a) through (e), and 516-6(a) through (d). These requirements are extracted material from NFPA 33-1995 and 34-1995 and are very specific. These standards and the *NEC®* sections referenced must be followed precisely.

Wiring and Equipment above Class I and Class II Locations

In these locations, all fixed wiring must be in metal raceways, rigid nonmetallic conduits, electrical nonmetallic tubing, or in types MI, MC, TC, or SNM cable where the wiring method leads into or through the classified Class I or Class II location. Suitable seals must be employed. Equipment that might produce a spark or arc, such as lamps, lampholders, switches, motors, or other similar equipment, shall not be located above classified areas where fresh goods are handled unless of the totally enclosed type.

Grounding

All metal raceways, the metal armor or metallic sheath on cables, and all non-current-carrying metal parts of fixed or portable equipment must be grounded, regardless of voltage, and comply with the provisions of Article 250. Grounding in Class I locations shall comply with Section 501-16.

CHAPTER 12

Health Care Facilities, Residential and Custodial Facilities, and Limited Care Facilities

The requirements for these areas are found in *NEC*® Article 517. Additional information regarding the installation, testing, and safeguards of these areas can be found in NFPA 99-1996. Specific definitions of areas within these facilities can be found in *NEC*® Section 517-3.

Health care facilities include the electrical equipment installations in hospitals, nursing homes, extended care facilities, clinics, and dental offices. In addition, *NEC*® Article 517 includes provisions for electrical installations in areas provided for patients undergoing a type of treatment involving the use of electrical energized equipment connected to the heart. Requirements for essential and emergency electrical systems in hospitals have also been covered in this *NEC*® article. These specialized systems must meet exacting and detailed requirements, a discussion of which is beyond the scope of this manual.

The following comments are limited to the electrical requirements in special hazardous areas in hospitals, usually identified as an anesthetizing location, including hospital operating rooms, delivery rooms, anesthesia rooms, and any corridor, utility room, or other area that might be used for this purpose or where flammable anesthetics are stored. The hazardous space and flammable anesthetizing locations extend from the floor to a height of 5 feet and are classified as Class I, Division 1 locations. The entire room where flammable anesthetics are stored is classified as a Class I, Division 1 location. All fixed wiring and equipment and all portable equipment, including lamps and other utilization equipment operating at more than 10 volts between conductors, shall comply with the requirements of Article 501 for Class I, Division 1 locations. All such equipment shall be specifically approved for the hazardous atmospheres involved.

Receptacles and attachment plugs in hazardous (classified) location(s) shall be listed for use in Class I, Group C hazardous (classified) locations and must have provision for the connection of a grounding conductor. Receptacles and attachment plugs located above hazardous anesthetizing locations shall be listed for hospital use for services of prescribed voltage, frequency, rating, and number of conductors with provision for the connection of the grounding conductor. This requirement shall apply to attachment plugs and receptacles of the 2-pole, 3-wire grounding type for single-phase 120-volt, nominal, AC service.

Each power circuit within, or partially within, a flammable anesthetizing location as referred to in Section 517-60 shall be isolated from any distribution system supplying other than anesthetizing locations. Maximum potential on the primaries or secondaries of isolating transformers is limited to 300 volts. Some specific allowances in Section 517-61 might apply.

Branch circuits that supply an anesthetizing location are not permitted to supply any other location. Special insulation on circuit conductors is also required to minimize capacitance of circuitry. These circuits are required to be supplied by the ungrounded side of an isolated transformer. Branch circuits supplying only fixed lighting fixtures in nonhazardous areas other than surgical lighting fixtures or approved permanently installed X-ray equipment can be supplied by grounded circuits. If supplied by a grounded circuit, the conductors must be in a separate raceway from the conductors of the ungrounded circuit. The lighting fixtures and X-ray equipment must be located 8 feet above the floor, with switches controlling the grounded circuits located outside the anesthetizing location.

Low-voltage current circuits are required by the *NEC*® for electrical apparatus and equipment having exposed current-carrying parts that are frequently in contact with bodies of patients. Such equipment circuits must be designed and approved for operation of 8 volts or less. More detailed and complete information can be found in NFPA 99. **Note:** The NFPA 99 handbook will provide great assistance when designing or installing a health care facility.

CHAPTER 13

Intrinsically Safe Systems

What Is Intrinsic Safety?

Intrinsic safety is an explosion-prevention design technique applied to electrical equipment and the wiring of hazardous locations where flammable or combustible material is present. The technique is based on limiting electrical and thermal energy levels below that which is required to ignite the specific hazardous atmospheric mixture present. Intrinsically safe wiring shall not be capable of releasing sufficient electrical or thermal energy under normal or abnormal conditions to cause ignition to a specific flammable or combustible atmospheric mixture in its most easily ignitable concentration. It is important to define the class and group for which any proposed intrinsically safe electrical circuits are to be installed. Because intrinsic safety is a technique for the worst-case hazardous locations, consideration of the division is not necessary. Because intrinsic safety maintains energy levels below that required to ignite specific hazardous mixtures, it is important to know what the energy allowances are for operational and safety considerations. Intrinsically safe systems, Article 504, was introduced as an acceptable wiring system in the 1990 *NEC*®. Careful interpretation and application of this article will provide safe installations that are free from hazard.

Definition

Article 504 covers the installation of intrinsically safe apparatus, wiring, and systems for Class I, II, and III locations. The specific and unique definitions of terms for this article appear in Section 504-2.

The use of intrinsically safe equipment is primarily limited to process control instrumentation. Because these electrical systems lend themselves to the low-energy requirements, ANSI/UL 913 provides information on the design test conditions and evaluation of safe apparatus and associated apparatus for use in Class I, Class II, and Class III, Division 1 hazardous classified locations. Underwriters Laboratories Inc. and Factory Mutual list several devices in this category.

The equipment and associated wiring must be installed so they are positively separated from the nonintrinsically safe circuits. It is very important that these systems be designed and installed precisely as per the control drawings. It is also very important that all equipment be marked with the control drawing number. Following these guidelines will help prevent corruption of the integrity of the system. Induced voltages could defeat the concept of intrinsically safe systems.

NEC® Section 504-1 clarifies that this article is intended to cover only intrinsically safe apparatus and wiring in Class I, II, and III locations and not intrinsically safe systems that are not located in a classified area. All intrinsically safe apparatus and associated apparatus must be approved and accept as modified in Article 504. All applicable articles of the Code are applied to these systems. It is important to note that in applying *NEC®* Article 504, only the entire intrinsically safe system must be installed in accordance with control drawings, as required in *NEC®* Section 504-10. The control drawing identification must be marked on all equipment and apparatus. Generally, intrinsically safe systems are permitted using any wiring method suitable for unclassified locations. However, sealing shall be provided in accordance with *NEC®* Section 504-70, and separation of intrinsically safe systems shall be provided in accordance with *NEC®* Section 504-30.

Wiring Methods

General wiring methods are permitted. This includes any of the wiring methods in Chapter 3 that are appropriate for the location and use. **Warning:** Corrosion and physical protection have not been considered, only circuit safety. It is the designer's responsibility to consider all facets of the installation for performance and safety.

It should be noted that intrinsically safe systems are designed to always fail in the safe position. Should a wire or cable be crushed or cut accidentally, the operation of the equipment must be fail-safe in every instance. No arcing or sparking will occur because of the design and nature of intrinsically safe systems. Designers generally require physical protection, such as metal raceways, to protect intrinsically safe conductors and to provide assurance from costly accidental shutdowns and operational stoppages.

Sealing

NEC® Section 504-70 requires that conduits and cables be sealed in accordance with the hazardous location in which they are installed, such as Article 501 for Class I hazardous flammable vapors and gases. Sealing of intrinsically safe systems must be provided as specified in Section 501-5. For flammable or combustible dust, sealing is required as specified in *NEC®* Section 502-5. Seals are not required for enclosures that contain only intrinsically safe apparatus.

Identification

Permanent identification "intrinsic safety wiring" is required at all terminals and throughout the entire length of the wiring method and support system at intervals no more than 25 feet apart. This requirement applies within the hazardous area and in the nonhazardous locations. Industry practice has been to identify this system with the color light blue for the cable systems; however, color coding is not required. Color coding is permitted where no other conductors used are light blue in color. For specific requirements, see *NEC®* Section 504-80(a) and (b). This strict marking identification is necessary in both hazardous and nonhazardous locations to prevent inadvertently mixing other systems with intrinsically safe systems both during the initial installation and at a later date.

Electrical and Mechanical Separation

Open wiring of conductors and cables is permitted but must be separated from nonintrinsically safe circuits at least 1.97 inches (50 mm) and secured. It must also be secured and separated a minimum of 1.97 inches (50 mm) from conductors or cables of other systems not intrinsically safe. Intrinsically safe conductors generally must not be placed in the same raceway, cable tray, or cable with circuits of other nonintrinsically safe systems. Exceptions to this general rule can be found in *NEC®* Section 504-30(a)(1) and (2).

Within enclosures, conductors must be secured and separated a minimum of 1.97 inches (50 mm) from conductors or cables of other systems not intrinsically safe. Separate wiring compartments or physical barriers are preferred by most designers.

Intrinsically safe circuits shall be separated from other intrinsically safe circuits by being placed in separate cables or with a grounded metal shield or by a minimum of 0.01 inches of insulation.

Grounding

All intrinsically safe systems associated apparatus, enclosures, raceways, cables, and cable shields must be grounded. The control drawings might also require supplementary bonding to the grounding electrode to some associated apparatus, such as intrinsic safety barriers.

Safety Assured

There are a number of ways to design systems to ensure intrinsic safety. However, all provide separation by an electrical barrier (associated apparatus) that separates intrinsically safe protected wiring from the unprotected nonintrinsically safe wiring. In all cases, these barriers will limit energy into the hazardous location under unique specified fault conditions. The barriers generally fall into two basic categories: isolated and nonisolated.

Intrinsic safety barriers are passive networks consisting of voltage-limiting diodes and current-limiting resistors or semiconductive current-limiting components. They allow the electrical signal to flow while limiting the energy level. These are usually protected from opening a high-fault condition by a fuse. To provide overvoltage protection, intrinsic safety barriers must be grounded in accordance with the control drawings and must be located outside the classified hazardous area.

Intrinsic safety galvanic isolation devices provide isolation between intrinsic safe wiring and nonintrinsic safe wiring without the necessity of a ground connection for overvoltage. These are generally equipped with signaling condition devices. Although this new accepted installation system will provide safe installations, it is paramount that they be installed precisely as designed and shown on the control drawings as permitted in *NEC®* Article 504. Manufacturers of these systems should provide additional installation and startup guidance to the installer.

Wiring Methods
for Hazardous Locations

Before selecting a wiring method, many factors must be considered: environment, physical protection requirements, temperature, and all other pertinent conditions that might affect the material selected. Once this has been done, a wiring method can be selected from Chapter 3 of *NEC*®. Each installation generally requires several wiring methods to meet all the conditions, such as a method that is acceptable for "dry" concealed locations or a method that provides "physical protection." It might also be necessary to select a "flexible" wiring method for equipment connection and a material that can be used in "wet locations" and "underground." There are wiring methods that are acceptable in all locations. The final choice must comply with the requirements of *NEC*® Sections 110-3(a) and (b) and is generally made by the designer. Once the wiring method or methods have been selected, it is the responsibility of the installer to comply with the provisions in *NEC*® Article 300 and the *NEC*® article that specifically covers each wiring method chosen before making the installation (see Figure 14–1). **Note:** General wiring methods are all found in the *NEC*®, Chapter 3. However, when wiring a hazardous location, the provisions of *NEC*® Chapter 5 generally limit the number of different types of wiring methods.

Protection against Physical Damage

The function of an electrical raceway is to facilitate the insertion and extraction of the conductors and to protect them from mechanical injury. All raceway types facilitate easy insertion and extraction of conductors. However, only rigid metal conduit, intermediate metal conduit, and rigid nonmetallic conduit Schedule 80 provide protection of conductors against physical damage (see Figure 14–2). Electrical metallic tubing provides physical protection, except where severe physical damage to the conductors is likely.

The designer and installer can no longer rely on all raceways to provide physical protection. Steel raceways, rigid metal conduits as covered in *NEC*® Article 346, and intermediate metal conduit as covered in *NEC*® Article 345 provide the greatest physical protection and are permitted in all applications of the code (see Figures 14–3 and 14–4). Schedule 80 PVC as covered in *NEC*® Article 347 does provide excellent physical protection. However, a fine-print note in *NEC*® Section 347-2 warns us that in cold temperatures PVC can become brittle and crack or break under impact. Extreme cold can cause some

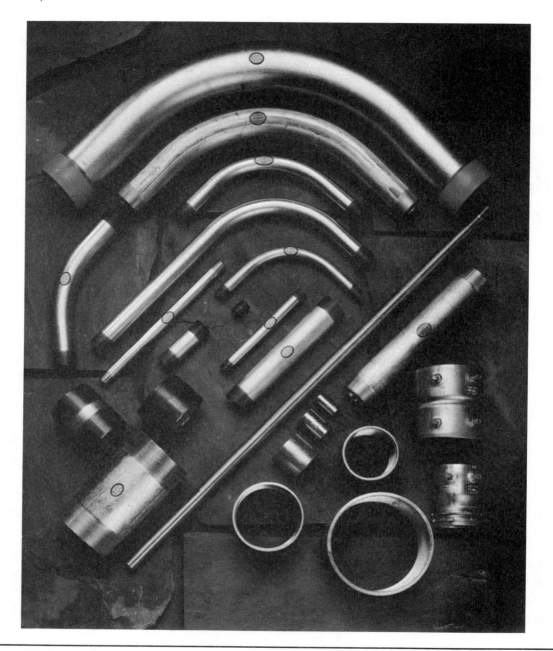

Figure 14–1 Conduit nipples, elbows, and fittings. (Courtesy of Picoma Industries Inc.)

Figure 14–2 Typical construction project installing steel raceways that provide excellent physical protection that is needed in a concrete slab during installation pour. (Courtesy of Allied Tube & Conduit.)

Figure 14–3 A typical construction project installing conduit in a concrete slab; this is an excellent example of how physical damage could occur during a concrete pour. (Courtesy of Allied Tube & Conduit.)

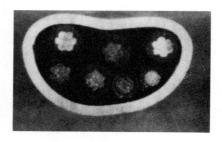

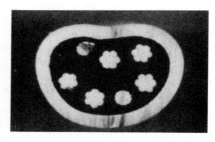

Figure 14–4 Test result summary. The after-impact internal diameter exceeded the 40% allowable fill in all tests for both rigid steel and IMC conduit. Conductors were not damaged in either rigid steel or IMC, and in all cases conductors were easily extracted. (Courtesy of Allies Tube & Conduit.)

nonmetallic conduits to become brittle and therefore more susceptible to damage from physical contact. Therefore, Schedule 80 can be used for physical protection, but consideration must be given to the environment in which it is to be installed as well as the physical protection needed under those conditions. Rigid nonmetallic conduit is permitted to be used only in hazardous locations in very limited and specific locations. Electrical metallic tubing as covered in *NEC*® Article 348 provides great physical protection for all but severe physical damage locations. EMT, like nonmetallic conduit, is permitted only on a limited basis in hazardous locations.

Raceways are known to frequently outlast repeated reconstruction and remodeling of the buildings they serve, and all these things must be considered when the designer is selecting the raceway for the specific installation. Great care is needed in making these decisions because one of the important contributing factors in selecting a raceway is the long-term economy of the installation. In the case of electrical modernization, old wires can often be pulled out and new wires pulled in without structural remodeling. It is up to the designer to select a raceway type that will provide this feature. *NEC*® Section 300-4 requires conductors to be adequately protected where subject to physical damage in both exposed and concealed locations, except for intermediate and metal conduit, rigid metal conduit, rigid nonmetallic conduit, and electrical metallic tubing. Raceways and cables shall be protected where run through bored holes or notches in wood, through metal framing members, and where run parallel to the framing members where a distance of 1¼ inches from the edge of the framing member cannot be maintained. The raceway or cable type shall be protected from penetration by screws or nails with a steel plate sleeve or equivalent of at least ⅟₁₆ inch thick. It is the responsibility of the designer to select a raceway type that will provide protection to the conductor in the environment for which it is installed. The degree of physical damage that the wiring method might be subjected to should dictate the raceway type to be selected. For above-ground installations over 600 volts nominal, the conductors shall be protected by installing them in rigid metal conduit, intermediate metal conduit, or rigid nonmetallic conduit. Conductors can also be run in cable trays, bus ways, or cable bus in some installations. The installation requirements for installations over 600 volts are found in *NEC®* Article 300. Section 300-37(a) governs above-ground installations, and Section 300-50 provides the physical protection requirements for underground installations over 600 volts where they emerge from the ground.

Although raceways were originally designed to protect conductors from mechanical injury and to provide the versatility for future changes by facilitating the insertion and extraction of conductors, in the past thirty years specialty raceways have been developed, accepted, and included in the *NEC®*.

Underground Requirements

NEC® Sections 300-5 and 300-50 covers the requirements for underground installations. *NEC®* Tables 300-5 and 300-50 gives the minimum cover requirements for cable or conduit to be installed (see Figures 14–5 and 14–6). All underground installations shall be grounded and bonded in accordance with *NEC®* Article 250. Where approved for direct burial, cable installed under a building shall be in a raceway that must extend beyond the outside wall of the building. Direct-buried cables and cables emerging from the ground shall be protected from physical protection in a raceway extending from the minimum cover distance required in *NEC®* Section 300–5 below grade to a point at least 8 feet above finish grade. Conductors entering a building shall be protected to the point of entrance. Where a raceway or enclosure is subject to physical damage, the conductors must be installed in rigid metal conduit, intermediate metal conduit, or a Schedule 80 rigid nonmetallic conduit or equivalent. Raceways or cables shall not be backfilled with a material that will damage, prevent adequate compacting, or contribute to corrosion of raceways or cables. When it is necessary to prevent physical damage from occurring to a raceway or cable, the raceway or cable can be covered with select material (sand), suitable running boards, sleeves, or other approved means. Raceways through which moisture might contact energized live parts, such as conductors entering the building from outside into panelboards or switchboards that might contact live parts, shall be sealed or plugged at either or both ends. If the presence of hazardous gases or vapors are present, this might also require the sealing of underground raceways where they enter the building. A bushing or terminal fitting shall be placed on the end of a conduit or raceway that terminates underground where cables approved for direct burial, such as type "UF," emerge as a direct-burial wiring method. A seal incorporating the physical protection characteristics of a bushing is permitted in lieu of the bushing. *NEC®* Article 300, Part B has similar requirements for underground installations of over 600 volts nominal (Figure 14–6).

Exposure to Different Temperatures

Raceways exposed to different temperatures are covered in *NEC®* Section 300-7. This section of the *NEC®* requires that where portions of interior raceways are exposed to widely different temperatures, such as in refrigeration or cold storage plants, circulation of air from the warmer to the colder section through a raceway must be prevented by sealing the open ends of the wiring method where the temperature change occurs. This is to reduce condensation from building up in the equipment being served by that raceway. Although the *NEC®* requires this precautionary measure only for interior installations, it

TABLE 300-5. Minimum Cover Requirements, 0 to 600 Volts, Nominal, Burial in Inches (Cover is defined as the shortest distance measured between a point on the top surface of any direct buried conductor, cable, conduit, or other raceway and the top surface of finished grade, concrete, or similar cover.)

Location of Wiring Method or Circuit	Type of Wiring Method or Circuit				
	1 Direct Burial Cables or Conductors	2 Rigid Metal Conduit or Intermediate Metal Conduit	3 Nonmetallic Raceways Listed for Direct Burial without Concrete Encasement or Other Approved Raceways	4 Residential Branch Circuits Rated 120 Volts or Less with GFCI Protection and Maximum Overcurrent Protection of 20 Amperes	5 Circuits for Control of Irrigation and Landscape Lighting Limited to Not More than 30 Volts and Installed with Type UF or in Other Identified Cable or Raceway
All Locations Not Specified Below	24	6	18	12	6
In Trench Below 2-Inches Thick Concrete or Equivalent	18	6	12	6	6
Under a Building	0 (In Raceway Only)	0	0	0 (In Raceway Only)	0 (In Raceway Only)
Under Minimum of 4-Inch Thick Concrete Exterior Slab with No Vehicular Traffic and the Slab Extending Not Less than 6 Inches Beyond the Underground Installation	18	4	4	6 (Direct Burial) 4 (In Raceway)	6 (Direct Burial) 4 (In Raceway)
Under Streets, Highways, Roads, Alleys, Driveways, and Parking Lots	24	24	24	24	24
One- and Two-Family Dwelling Driveways and Outdoor Parking Areas, and Used Only for Dwelling-Related Purposes	18	18	18	12	18
In or Under Airport Runways, Including Adjacent Areas Where Trespassing Prohibited	18	18	18	18	18

Note 1. For SI units: 1 inch = 25.4 mm.
Note 2. Raceways approved for burial only where concrete encased shall require concrete envelope not less than 2 inches thick.
Note 3. Lesser depths shall be permitted where cables and conductors rise for terminations or splices or where access is otherwise required.
Note 4. Where one of the wiring method types listed in columns 1–3 is used for one of the circuit types in columns 4 and 5, the shallower depth of burial shall be permitted.
Note 5. Where solid rock is encountered, all wiring shall be installed in metal or nonmetallic raceway permitted for direct burial. The raceways shall be covered by a minimum of 2 inches of concrete extending down to rock.

Figure 14–5 Burial depth requirements for installations of 600 volts and less. (Reprinted with permission from NFPA 70-1999, *National Electrical Code®*, Copyright © 1998, National Fire Protection Association, Quincy, MA 02269.)

TABLE 300-50. Minimum Cover Requirements (Cover is defined as the shortest distance in inches measured between a point on the top surface of any direct buried conductor, cable, conduit, or other raceway and the top surface of finished grade, concrete, or similar cover.)

Circuit Voltage	Direct Buried Cables	Rigid Nonmetallic Conduit Approved for Direct Burial*	Rigid Metal Conduit and Intermediate Metal Conduit
Over 600–22kV	30	18	6
Over 22kV–40kV	36	24	6
Over 40kV	42	30	6

For SI units: 1 inch = 25.4 mm.
*Listed by a qualified testing agency as suitable for direct burial without encasement. All other nonmetallic systems shall require 2 inches (50.8 mm) of concrete or equivalent above conduit in addition to above depth.

Exception No. 1: Areas subject to vehicular traffic, such as thoroughfares or commercial parking areas, shall have a minimum cover of 24 inches (610 mm).

Exception No. 2: The minimum cover requirements for other than rigid metal conduit and intermediate metal conduit shall be permitted to be reduced 6 inches (152 mm) for each 2 inches (50.8 mm) of concrete or equivalent protection placed in the trench over the underground installation.

Exception No. 3: The minimum cover requirements shall not apply to conduits or other raceways that are located under a building or exterior concrete slab not less than 4 inches (102 mm) in thickness and extending not less than 6 inches (152 mm) beyond the underground installation. A warning ribbon or other effective means suitable for the conditions shall be placed above the underground installation.

Exception No. 4: Lesser depths shall be permitted where cables and conductors rise for terminations or splices or where access is otherwise required.

Exception No. 5: In airport runways, including adjacent defined areas where trespass is prohibited, cable shall be permitted to be buried not less than 18 inches (457 mm) deep and without raceways, concrete enclosement, or equivalent.

Exception No. 6: Raceways installed in solid rock shall be permitted to be buried at lesser depth where covered by 2 inches (50.8 mm) of concrete, which shall be permitted to extend to the rock surface.

Figure 14–6 Burial depth requirements for installations of over 600 volts. Note: Local utility or the AHJ requirements might exceed these minimum requirements. (Reprinted with permission from NFPA 70-1999, *National Electrical Code®*, Copyright © 1998, National Fire Protection Association, Quincy, MA 02269.)

is also recommended where wiring methods pass from outside to environmentally controlled areas inside a building (see Figure 14–7). Good installation practices will dictate entry into equipment in a manner so that the condensate will not drip on exposed live parts. It is not uncommon for improper installations to short out, causing expensive downtime when this occurs. Special fittings designed for sealing conduit openings are not required for this application. Field methods for sealing can be accomplished using "duct-seal" or other removable putty that will not deteriorate conductor insulation. Expansion joints are required when necessary to compensate for thermal expansion and contraction. The specific requirements for nonmetallic raceways, which have a higher coefficient of expansion, are covered below (Figure 14–8).

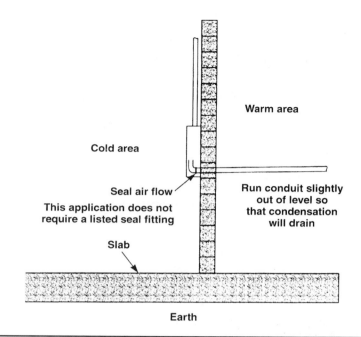

Figure 14–7 Seals to prevent moisture from condensation as required per *NEC®* Section 300-7 where conduit systems pass from areas with widely different temperatures. This section does not require an approved hazardous area seal fitting.

TABLE 347-9 Expansion Characteristics of PVC Rigid Nonmetallic Conduit (Coefficient of Thermal Expansion = 3.38×10^{-5} in./m./°F.)

Temperature Change in Degrees F	Length Change in Inches per 100 Feet of PVC Conduit	Temperature Change in Degrees F	Length Change in Inches per 100 Feet of PVC Conduit	Temperature Change in Degrees F	Length Change in Inches per 100 Feet of PVC Conduit	Temperature Change in Degrees F	Length Change in Inches per 100 Feet of PVC Conduit
5	0.2	55	2.2	105	4.2	155	6.3
10	0.4	60	2.4	110	4.5	160	6.5
15	0.6	65	2.6	115	4.7	165	6.7
20	0.8	70	2.8	120	4.9	170	6.9
25	1.0	75	3.0	125	5.1	175	7.1
30	1.2	80	3.2	130	5.3	180	7.3
35	1.4	85	3.4	135	5.5	185	7.5
40	1.6	90	3.6	140	5.7	190	7.7
45	1.8	95	3.8	145	5.9	195	7.9
50	2.0	100	4.1	150	6.1	200	8.1

Figure 14–8 *NEC®* Table 347-9. Rule of thumb: To utilize this table for metal raceways, multiply the calculated length change in the table by 0.10 for aluminum and by 0.05 for steel raceways. (Reprinted with permission from NFPA 70–1999, *National Electrical Code®*, Copyright © 1998, National Fire Protection Association, Quincy, MA 02269.)

Expansion Properties of Raceways

Like all construction materials, raceway systems will expand and contract with variations in temperatures. The coefficient of linear expansion for various conduits is as follows:

- PVC conduit 3.38×10^5 in./in./°F
- Aluminum conduit 1.2×10^5 in./in./°F
- Steel conduit 6.50×10^{-6}/0.0000065 inches per inch of conduit for each °F in temperature change $\times 10^5$ in./in./°F

Note: The coefficient of expansion is for the material; therefore, the formulas apply to all types of PVC raceway products. All types of steel raceways and all types of aluminum raceway products are covered by the respective applicable formula. Expansion fittings are not generally required except in extremely long runs where exposed to a high degree of temperature change, such as rooftops.

The designer and installer generally need not be concerned with thermal expansion and contraction in steel and aluminum installations because of the low coefficient of linear expansion. *NEC*® Section 347-9 states that an expansion coupling is needed wherever the change in length due to temperature variation will exceed .25 inches. *NEC*® Section 300-7(b) states, "Raceways shall be provided with expansion joints where necessary to compensate for thermal expansion and contraction." Expansion must be a consideration in all raceways, both metallic and nonmetallic. However, because of the low coefficient of expansion in steel raceways, only about 5% of (PVC) nonmetallic rigid conduit expansion joints are rarely necessary to comply with *NEC*® Section 300-7(b). However, expansion joints are often necessary in all raceways types, both metallic and nonmetallic and ferrous and nonferrous for architectural considerations. Where electrical raceway systems cross building expansion joints, which are commonly found in large buildings, the electrical raceway system should include an expansion joint to permit the raceway to move with the building. Other considerations for expansion might be encountered where buildings are designed as earthquake-proof, such as commonly found in the West Coast areas. Where these flexible architectural joints are encountered, the electrical raceway system should also include expansion joints to compensate for any movement encountered during an earthquake (see Figures 14–8 and 14–9).

However, nonmetallic raceways, because of the high coefficient of expansion, have special requirements in *NEC*® Article 347. *NEC*® Section 347-9 states that expansion joints for rigid nonmetallic conduit shall be provided to compensate for thermal expansion and contraction except where less than .25 inches of expansion is encountered. This new exception in the 1996 *NEC*® clarifies that all installations of rigid nonmetallic conduit encountering an expansion of over .25 inches is required to have an expansion joint installed.

NEC® Chapter 9, Table 10 provides information to the designer and installer as to when expansion is needed (see Figure 14–8). A good rule of thumb in utilizing this table for other than PVC rigid nonmetallic conduit is to apply a multiplier for other materials. A multiplier of .05 would result in the projected expansion of ferrous (steel) metal conduit, or about 1/20 of PVC. A multiplier of .1 would result in the projected expansion of (aluminum) nonferrous rigid metal conduit, or about 1/10 of PVC. To clarify, see the example in Figure 14–9.

A manufacturer of nonmetallic conduit recommends that 30°F be added to the estimated temperature range when PVC conduit is installed in direct sunlight to allow for radiant heating. This is due to the gray color of the raceway, which absorbs that radiant heating.

An expansion coupling consists of two sections of conduit, one telescoping inside the other. When installing expansion couplings, alignment of the piston and the barrel is

Example

380 ft. of conduit is to be installed on the outside of a building exposed to the sun in a single straight run. It is expected that the conduit will vary in temperature from 0°F in the winter to 140°F in the summer (this includes the 30°F for radiant heating from the sun). The installation is to be made at a conduit temperature of 90°F. From the table, a 140°F temperature change will cause a 5.7 in. length change in 100 ft. of conduit. The total change for this example is 5.7″ × 3.8 = 21.67″ which should be rounded to 22″. The number of expansion couplings will be 22 ÷ coupling range (6″ for E945, 2″ for E955). If the E945 coupling is used, the number will be 22 ÷ 6 = 3.67 which should be rounded to 4. The coupling should be placed at 95 ft. intervals (380 ÷ 4).

The proper piston setting at the time of installation is calculated as explained above.

$$0 = \left[\frac{140 - 90}{140} \right] 6.0 = 2.1 \text{ in.}$$

Insert the piston into the barrel to the maximum depth. Place a mark on the piston at the end of the barrel. To properly set the piston, pull the piston out of the barrel to correspond to the 2.1 in. calculated above.

See the drawing below.

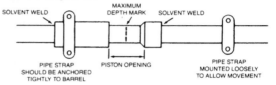

Figure 14-9

Figure 14–9 The above calculations are for rigid nonmetallic conduit. For steel raceways, multiply the calculated length change of 21.7 inches by 0.05, which equals 1.08 inches of expansion. For aluminum raceways, multiply the calculated length change of 21.7 inches by 0.10, which equals 2.17 inches of expansion. (Courtesy of Carlon, a Lampson Sessions Company.)

important. Be sure to mount the expansion joint level or plumb for best performance. For a vertical run, the expansion coupling must be installed close to the top of the run with the barrel portion of the expansion fitting pointing down so that rainwater does not run into the opening. The lower end of the conduit run must be secured at the bottom so that any length change due to temperature variation will result in an upward movement (see Figure 14–10).

Example

Three hundred and eighty feet of nonmetallic conduit is to be installed on the outside of a building exposed to the sun in a single straight run. It is expected that the conduit will vary in temperature from 0°F in the winter to 100°F in the summer. We must add 30°F for radiant heating from the sun for a total of 140°F. The installation is to be made at a conduit temperature of 90°F.

- From the table, a 140°F temperature change will cause a 5.7-inch length change for each 100 feet of conduit.

Determine The Piston Opening

The expansion joint must be installed to allow both expansion and contraction of the conduit run. The correct piston opening for any installation condition should use the following formula:

$$0 = \left[\frac{T \text{ max} - T \text{ installed}}{\Delta T} \right] E$$

where:
0 = piston opening (in.)
T max = maximum anticipated temperature of conduit (°F)
T ins = temperature of conduit at time of installation (°F)
ΔT = total change in temperature of conduit (°F)
E = expansion allowance built into each expansion coupling (in.)

STANDARD EXPANSION COUPLINGS

E945 series expansion couplings are designed to compensate for length changes due to temperature variations in exposed conduit runs. See explanation on previous page.

NOTE: Do not use expansion coupling when encased in concrete. Conduit is immobilized by the concrete and will conform to the expansion rate of the concrete.

Expansion Couplings

| Part No. | Size | Ctn. Qty. | Lay Lengths | | Available Length Expansion-Contraction |
			Stop to Stop Total Closed	Stop to Stop Total Open	
E945D	½	50	12¼	18⅝	6
E945E	¾	50	12¼	18⅝	6
E945F	1	45	12¾	19⅛	6
E945G	1¼	30	12¾	19⅛	6
E945H	1½	25	12¾	19⅛	6
E945J	2	15	13½	19⅞	6
E945K	2½	10	14	20⅜	6
E945L	3	10	16½	23	6
E945M	3½	5	16½	23	6
E945N	4	5	17½	24	6
E945P	5	3	18½	24½	6
E945R	6	2	20½	26½	6

SHORT EXPANSION COUPLINGS

(Expands to a maximum of 2")

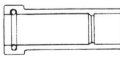

Part No.	Size	Ctn. Qty.
E955D	½	40
E955E	¾	40
E955F	1	25
E955G	1¼	15
E955H	1½	10
E955J	2	6

Figure 14–10   (Courtesy of Carlon, a Lampson Sessions Company.)

- Thus, the total change for this example is 5.7 inches $\times$ 3.8 = 21.67 inches (which should be rounded up to 22 inches).
- Expansion couplings come in two types; one has a range of 2 inches, the other a range of 6 inches.
- The number of expansion couplings will be 22 divided by the coupling range (6 inches or 2 inches). *NEC®* 300-15(c) requires that all fittings and connectors be used only with specific wiring methods for which they are designed and listed.
- If the 6-inch expansion coupling is used, the number will be 22 ÷ 6 = 3.67 (which must be rounded up to 4 expansion couplings).
- The couplings should be placed at 95-foot intervals (380 feet ÷ 4).
- The proper piston setting at the time of installation is calculated as explained in Figure 14–10. Formula X, 0 = (140 − 90/140) 6.0 = 2.1 inches.
- Insert the piston into the barrel to the maximum depth; place a mark on the piston at the end of the barrel to properly set the piston; pull the piston out of the barrel to correspond to the 2.1 inches calculated above.
- If this same 380-foot run conduit were installed using steel conduit, the expansion would be 21.67 × .05, or approximately 1.1 inches. If run in aluminum conduit, expansion would be 21.67 × .1, or approximately 2.2 inches.

Summary

1. Anticipate expansion and contraction of PVC conduit in all above-ground exposed installations.
2. Use an expansion coupling when the length change due to temperature variation will exceed .25 inches.
3. PVC conduit expands 4.1 inches for each 100 feet run and 100°F of temperature change.
4. Steel conduit expands approximately .2 inches per each 100-foot run and 100°F temperature change.
5. Aluminum conduit would expand approximately .4 inches for each 100-foot run and 100°F temperature change.
6. Align expansion coupling with conduit run to prevent binding (plumb or level).
7. Follow the instructions to set the piston opening.
8. Rigidly fix the outer barrel of the expansion coupling so it cannot move.
9. Mount the conduit connected to the piston loose enough to allow the conduit to move as the temperature changes (see Figures 14–8, 14–9, and 14–10.)

Note: Manufacturers recommend that for all above-ground installations where temperature change in excess of 25°F (15°C) is anticipated, expansion joints shall be installed. See *NEC®*, Chapter 9, Table 10 for expansion characteristics.

Secure and Support Requirements

Although the specific requirements for securing and supporting each wiring method type is covered in the article governing that wiring method, general securing and supporting requirements are covered in Article 300. (All support methods are made up from hardware

Figure 14–11 Rigid junction box support and individual raceway support that must be approved by the AHJ. (Courtesy of Allied Tube & Conduit.)

items that are not "listed" by an approved testing laboratory.) The integrity of support methods must be field evaluated (see Figure 14–11). Careful consideration should be given to the methods used, as the integrity of the wiring method can be jeopardized by the use of inadequate support methods. *NEC*® Section 300-11 requires that all raceways, cable assemblies, boxes, cabinets, and fittings be supported and securely fastened in place. Ceiling support wires are generally not permitted as the sole support. This *NEC*® section permits branch circuit raceways to be supported on support wires where the equipment is supported by or located below the suspended ceiling but are not permitted to be used when used for the support of fire-rated ceilings. It is important to remember that Section 300-11 covers only branch circuits. *NEC*® Sections 725-6, 760-7, 770-7, 800-5, and 820-5 require access to electrical equipment behind panels, including suspended ceilings that are designed to allow access. These sections further state that access shall not be denied by an accumulation of wires and cables that prevent access. **Caution:** The designer should coordinate this support system with the architect designing the ceiling grid system.

Building inspectors might not permit the use of these support wires and might require independent support of raceway systems. Raceways are not to be used as a means of support for other raceways, cables, or nonelectrical equipment, except where identified for the purpose. Good support methods are important in maintaining the integrity of the raceway system. Generally, each wiring component must be independently supported. As an example of supporting requirements, each component part of the system must be independently supported and in compliance with Article 300 generally:

- The outlet or junction box must be supported in accordance with *NEC*® Article 370.
- The wiring method must be supported in accordance with the article governing it, such as electrical metallic tubing (EMT) Article 348 (see Figure 6–3 in Chapter 6 and Figure 7–4 in Chapter 7).
- The lighting fixtures must be in accordance with Article 410 (see Figure 14–12).

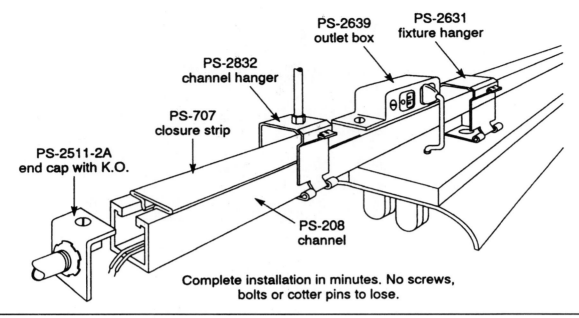

PS-2639
outlet box

PS-2631
fixture hanger

PS-2832
channel hanger

PS-707
closure strip

PS-2511-2A
end cap with K.O.

PS-208
channel

Complete installation in minutes. No screws, bolts or cotter pins to lose.

Figure 14–12 Strut can serve as a raceway in accordance with Article 352, Part C, as a support, or as both. (Courtesy of Allied Tube & Conduit.)

Mechanical and Electrical Continuity

Metal or nonmetallic raceways, cables, armors, and cable sheaths shall be continuous between cabinets, boxes, fittings, and other enclosures or outlets. There is an exception to this requirement in Section 300-12. The electrical continuity of conductors in raceways shall be continuous, and there shall be no splice or tap within the raceway itself. Again, there are some exceptions to this rule where it is desirable to allow splices and taps within the raceway, fitting, or box. These include auxiliary gutters, wireways, boxes, fittings, surface metal raceways, and bus ways. **Note:** Wrench-tight fittings and couplings and the proper supporting of the wiring method will ensure the compliance of these requirements.

Induction Heating

When an alternating current is flowing, an electromagnetic field exits around each energized conductor. It varies in strength as the current in the conductor varies from zero to maximum during each half-cycle. If the conductors in a circuit are in separate steel raceways, the changing electromagnetic field induces an electromotive force in each raceway. Current will then flow in each raceway. The magnitude of the current flow is determined by the impedance of the raceway as a current path. Current flow through this impedance can generate sufficient heat to raise the temperature of the raceway high enough to damage the conductor insulation. The current can also cause an additional voltage drop by inducing a back electromotive force in the conductor itself. One solution is to have a conductor of each phase and the neutral of the circuit and equipment-grounding conductor, where required, in each raceway or cable. The sum of the currents in one direction would equal the sum of the currents in the opposite direction at any instant, and the changing electromagnetic fields will balance by canceling each other out. Inductive heating will therefore be minimized (see Figure 14–13). *NEC*® Section 300-20 requires electrical circuits enclosed in metal raceways or cables to have each phase conductor, neutral, and

Induced Currents in Steel Raceways

NEC Sec. 300-20
Unbalanced load, 3-phase, 4-wire system

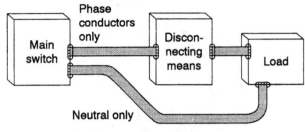

Entire loop subject to inductive heating
(violates *NEC Section 300-20*).

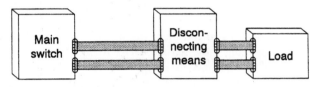

Neutral and phase conductors in each raceway.

Figure 14–13 Example of an installation that is in violation of Section 300-20(a) and (b) and will cause induced heating and fail (top); a correct installation (bottom). (Courtesy of American Iron & Steel Institute.)

equipment-grounding conductor, where used, installed in them. This same requirement applies to parallel conductor circuits carried over two or more parallel metal raceways or cables. Inductive heating can occur where the conductors of the circuit pass through the individual openings in the wall of the wireway or auxiliary gutter, metal pole box, or cabinet. When feeders are terminated at a panelboard, the inductive heating effect can be minimized by cutting slots in the metal between the conductor openings. Another method is to cover a rectangular opening cut in the metal cabinet with an insulated block containing individual openings for separate conductors. One or the other of these methods is required by *NEC*® Section 300-20.

There is an exception to this requirement where current is negligible, such as for secondary conductors of X-ray and electric discharge sign circuits. Aluminum is not a magnetic metal; therefore, heating due to hysteresis will not occur. However, induced current will be present. The heating will not be of sufficient magnitude to require the grouping of conductors or special treatment in passing the conductors through aluminum wall sections.

NEC® Section 300-3(b) also requires that all the conductors of the same circuit and, where used, the neutral and all equipment-grounding conductors be contained within the same raceway, cable tray, trench, cable, or cord, with some exceptions. *NEC*® Section 300-5(a) has similar requirements for direct-buried circuitry in a raceway or cable, and they shall be installed in close proximity in the same trench where there are direct-buried cables. Exception 2 of *NEC*® Section 300-5(i) permits isolated phase installations in nonmetallic raceways in close proximity where the conductors are paralleled as permitted in *NEC®* Section 310-4 and where the conditions of *NEC®* Section 300-20 are met. This is not a recommended practice, however. *NEC®* Section 310-4 makes very clear the importance of paralleling conductors to prevent imbalance loading or induction by stating that the conductors must be the same length, have the same conductor material, be the same size in circular mil area, have the same insulation type, and be terminated in the same manner.

Fine-print note (FPN) to *NEC®* Section 310-4 alerts us that the differences in inductive reactance and the unequal division of current can be minimized by the choice of materials, the methods of construction, and the orientation of the conductors. It is not the intent to require that the conductors of ungrounded (one-phase), neutral, or grounded circuit conductor be the same as those of another phase, neutral, or grounded circuit conductor to achieve balance (see Figure 14–13).

Resistance to Deterioration from Corrosion and Reagents

NEC® Section 110-3 requires the designer and installer to select the proper materials for each installation (see Figure 14–14). **Warning:** Chapter 5 of the *NEC®* does not consider the effects of deterioration from corrosion and reagents. It is the responsibility of the designer to consider the environment of the hazardous classified location. Guidelines for making these selections can be found in Underwriters Laboratories standards, the UL "Green Book," the *Electrical Constructions Materials Directory*, or the UL "White Book," which is the *General Information Directory for Electrical Construction, Hazardous Location and Electrical Heating and Air Conditioning Equipment.* In addition, *NEC®* Section 110-11 reminds us that deteriorating agents contribute to corrosion. Unless identified for use in the operating environment, no conductors or equipment shall be located in the environment. "Some cleaning and lubricating compounds can cause severe deterioration of many plastic materials used for insulating and structural applications in equipment." *NEC®*

Figure 14–14 Examples of PVC-coated galvanized steel rigid conduit, which offers excellent protection from all types of corrosion and physical protection. (Courtesy of Robroy Industries.)

Section 300-6 reminds us that metal raceways, armored cable, boxes, cable sheathing, cabinets, elbows, couplings, fittings, supports, and support hardware shall be of material suitable for the environment in which they are installed. Outer coatings should be of an approved corrosion resistant material, such as zinc, cadmium, or enamel (see Figures 14–15 and 14–16). Where protected from corrosion solely by enamel, they shall not be used outdoors or in wet locations as described below. Ferrous and nonferrous metal raceways, cable armor, boxes, cable sheathing, cabinets, elbows, couplings, fittings, supports, and support hardware shall be permitted to be installed in concrete or in direct contact with the earth and areas subject to severe corrosive influences when made of material judged suitable for the condition and when provided with corrosion protection approved for the condition (see Figure 14–14). In portions of dairies, laundries, canneries, and such locations where the walls are frequently washed or where surfaces are of absorbent materials such as damp paper or wood, the entire wiring system must be mounted so that there is at least ¼ inch of airspace separation from the wall or other mounting structure. In general, areas where acids and alkali chemicals are stored or might be present are highly corrosive areas, especially when these areas are wet or damp. Severe corrosive conditions might also be present in portions of meat-packing plants, tanneries, glue houses, stables, installations immediately adjacent to the seashore, and swimming pool areas. Areas where chemical de-icers are used and storage cellars or rooms for animal hides or casings, fertilizer, salt, and bulk chemicals are also to be considered highly corrosive.

NEC® Article 547 gives us some additional alerts when the installation falls within the scope of agricultural buildings and the concerns for corrosive atmospheres exist. These include poultry and animal facilities that might be damp from washing or sanitizing, where cleaning agents and similar conditions exist, livestock confinement areas, environmentally controlled poultry houses, and similar enclosed areas of an agricultural nature. With these sections, there is ample warning to the designer to make a correct and proper selection of the material to be specified. Since 1905, rigid metal conduit has been manufactured with a corrosion protection, originally a coating of black enamel, later by means of electrogalvanizing and hot-dip galvanizing. In 1961, the first PVC-coated rigid steel conduit was offered. Today there are numerous products on the market that provide specialty coatings for specific concerns relating to corrosion and deteriorating factors from the presence of severe environment and chemicals. Underwriters Laboratories, based on its tests and evaluations, state in the *General Information for Electrical Construction, Hazardous Location, and Electric Heating and Air Conditioning Equipment 1999* ("White Book") that the intermediate metal conduit, as covered by *NEC®* Article 345, and rigid metal conduit, as per *NEC®* Article 346, installed in concrete does not require supplementary corrosion protection. Rigid and IMC conduit installed in contact with soil does not generally require supplementary corrosion protection. EMT as covered in Article 348 does not generally require supplementary protection in concrete on grade or above but generally will if in concrete below grade or in direct contact with soil. Aluminum EMT in concrete or direct soil contact always requires supplementary protection. In the absence of specific local experience, soils producing severe corrosive effects are generally characterized by a low resistivity of less than 2,000 ohm-centimeters. Wherever ferrous metal conduit or tubing runs directly from concrete encasement to soil burial, severe corrosive effects are likely to occur on metal in contact with the soil. Supplementary nonmetallic coatings as part of the conduit have not been investigated for resistance to corrosion. Manufacturers should be contacted for this data. Supplementary nonmetallic coatings of thicknesses greater than .010 inches applied over metallic protective coatings are investigated with respect to flame propagation and

Rigid Steel Conduit, Intermediate Steel Conduit (IMC), and Electrical Metallic Tubing (EMT) are *Corrosion Resistant* for Wide Applicability

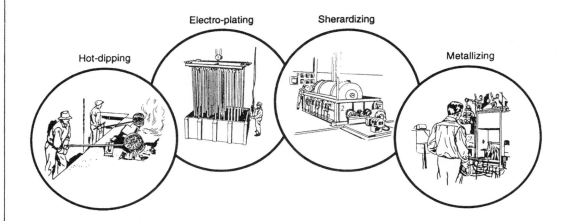

Both the exterior and the interior of tubular steel raceways are coated to protect from corrosion. Coatings are usually of zinc.

Zinc coating, or galvanizing, effectively seals the steel surfaces from corrosive moisture and other corrosive agents. In addition, the zinc coating may slowly sacrifice itself—and thus save the steel surfaces from corrosion—in those cases where the conduit may have been damaged and the steel exposed by a blow.

There are four methods of galvanizing steel conduit:

1. Hot-Dipping
2. Electroplating
3. Sherardizing
4. Metallizing

In the hop-dipping process, the clean tube is immersed in a molten zinc for a suitable time and then withdrawn. The tube is wiped with a superheated steam or air to remove any excess of zinc on either the exterior or interior, and to control the thickness of zinc.

In the electroplating process, the clean tube is placed in a solution of zinc salts and made a cathode of an electrical circuit so that the zinc is plated from the solution onto the outside of the tubing. Positive contact with the solution is obtained through zinc anodes, which replace the zinc content of the solution as it is used up.

In the sherardizing process, the conduit is first coated with zinc by heating a predetermined amount of commercially pure zinc powder and the steel pipe in a slowly revolving, sealed retort. A second method is also employed of electroplating a controlled thickness of zinc coating to the steel pipe. Both methods of zinc coating are subjected to accurately controlled furnace heat.

In the metallizing process, finely divided particles of zinc are metal-sprayed in a heated, semimolten condition onto the surface of a specially prepared tube to form an adherent coating.

Although hot-dipping and electroplating are the most common, any of the four methods of coating gives the tube a uniform coating of zinc of excellent corrosion resistance. (See A.S.A. for thickness of coating and test procedures.)

Figure 14–15

TABLE OF METALS IN GALVANIC SERIES

Corroded End (anodic or less noble)

MAGNESIUM
ZINC
ALUMINUM
CADMIUM
IRON OR STEEL
TIN
LEAD
NICKEL
BRASS
BRONZES
NICKEL-COPPER
ALLOYS COPPER

Any one of these metals and alloys will theoretically corrode while offering protection to any other that is lower in the series, so long as both are electrically connected.

As a practical matter, however, zinc is by far the most effective in this respect.

Because zinc is more active in galvanic couples than iron and steel, it provides these metals with electrolytic protection against rust. This protection is so effective that even though there be a small exposed area on the base metal, the attack of the elements will be directed to the zinc, and protection will continue as long as sufficient zinc remains.

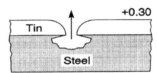

This is what happens at a small exposed area in a coating of a metal having a lesser tendency to go into solution than the base metal. The steel rusts away, protecting the tin.

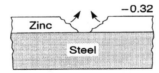

This is what happens at a small exposed area in a coating of a metal having a greater tendency to go into solution than the base metal. The zinc is consumed while protecting the steel from any attack.

Position in the galvanic series is not an infallible guide to galvanic protection because of other influences. Zinc is the one metal that will protect iron and steel under conditions of exposure to be expected.

Figure 14–16

detrimental effects to the basic corrosion protection provided by the protective coatings (see Figure 14–15). Many manufacturers today, in addition to the galvanized coating on metal raceways, also provide an additional layer of chromate or other organic coating to prevent formation of "white rust," which is the powdery substance that otherwise forms as a protective process of zinc (see Figure 14–15). This provides some additional protection that is not evaluated by UL. However, it certainly adds to the appearance as well as the protection on those raceway systems.

Many questions arise as to the suitability of the galvanized protective coating applied to rigid, IMC, and EMT conduit. The zinc coating need not be unscratched to remain effective. If a scratch should expose bare steel, moisture in the air will form an electrolyte on the surface of the raceway, and through galvanic action zinc will be transported to the exposed steel and deposited or plated out as an electrolytic band. This healing action of zinc affords protection even where wrench marks or scratches are as much as 0.1 inches wide. These metallic raceway types have provided excellent serviceability in countless installations over the past 100 years (see Figure 14–15). Corrosion is generally not a problem in most soil types. However, the final decision must be made by the designer and approved by the AHJ.

Rigid nonmetallic conduit has been perceived by many installers to be impervious to all corrosive atmospheres. This is not true! Rigid nonmetallic conduit, both Schedule 40 and Schedule 80, are generally acceptable for many environments; however, when making an installation in environments containing chemicals, check the manufacturer's recommendations (see Figure A–13 in the Appendix). If there are any questions for specific suitability in a given environment, prototype samples should be tested under actual conditions. Underwriters Laboratories' *General Information from Electrical Construction Materials Directory* ("White Book") states that listed PVC conduit is inherently resistant to atmospheres containing common industrial corrosive agents and will also withstand vapors or mist from caustic, pickling acids; plating bath; and hydrofluoric and chromic acids. PVC conduit, elbows, and bends, including couplings that have been investigated for direct exposure to other reagents, might be identified by the designation "Reagent Resistant" printed on the surface of the product. Such special uses are described as follows: "PVC conduit, elbows, and bends where exposed to the following reagents at 60°C or less, acetic, nitric (25°C only) acids, acids in concentrations not exceeding ½ normal; hydrochloric acid in concentrations not exceeding 30 percent; sulfuric acid in concentrations not exceeding 10 normal; sulfuric acid in concentrations not exceeding 80 percent (25°C only); concentrated or dilute ammonium hydroxide; sodium hydroxide solutions in concentrations not exceeding 50 percent; saturated or dilute sodium chloride solution; cottonseed oil, or ASTM #3 Petroleum Oil." An additional alternative was offered in 1961 when the first PVC-coated rigid steel was pioneered. Today there are several different types of coated conduit designed for different corrosive environments. These manufacturers offer a complete wiring method in special environments. They included coated conduit, fittings, and accessories and a complete package line, and they furnish experienced sales and engineering personnel capable of recommending the proper system for your application and providing a complete bill of material, including all conduit fittings, junction boxes, conduit bodies, and hardware for a complete system. Installation is made simple with standard rigid conduit tooling, and damage to the surface can easily be repaired by the installer at the time the damage occurs. The manufacturers of these products use standard listed galvanized rigid conduit in conformance with NC C80.1 and Underwriters Laboratories UL6 in accordance with *NEC*® Article 346 or listed intermediate metal conduit in compliance with UL 1242 and NC C80.6 in accordance with *NEC*® Article 345 listed products that is then coated to NEMA Standard RN-1. That material then has a protective coating 15 to 40 mils thick applied to the outside surface of the conduit. Protection is available on the outside, inside, or both in various mil thicknesses, according to the specific needs, and is suitable in all hazardous, heavy industrial, manufacturing, corrosive areas, such as bridges, highways, utility plants, substations. These products afford excellent protection from corrosion and reagents and deterioration from reagents as well as excellent physical protection and are ideal where both these conditions exist (see Figure 14–14). **Note:** See Interior and Exterior Chemical Resistance Charts in Figure A–15 in the Appendix. Today's designers and installers are offered an excellent selection of materials to meet all the needs of today's state-of-the-art installations. See also Figures A–13 and A–15 (manufacturer corrosion resistance data) in the Appendix.

Penetration of Fire-Resistive Assemblies

There are three major building codes that include many provisions, the purpose of which is to provide construction that not only will withstand the effects of fire in a portion of a building but also will reduce the likelihood of fire spreading within the building. For extremely large or high structures, building codes usually require fire-resistive construction as well as the use of noncombustible construction material. Building codes prescribe that in fire-resistive buildings, the columns, beams, girders, walls, and floor constructions must have a resistance to fire measured by a standard fire test procedure (ASTM E119, Standard Method of Fire Tests of Building Construction and Materials). The specific requirements of the assembly design under the classification system are found in Underwriters Laboratories' *Fire Resistive Directory 1999* ("Orange Book") and the *Building Materials Directory* ("Manila Book"). Other accredited labs also conduct tests to establish fire ratings. The three model building codes govern the fire-resistance requirements and stipulate the time-rated period for various types of occupancies (e.g., 1-hour or 2-hour). These requirements are found in the BOCA National Building Code Article 9, ICBO Uniform Building Code Chapter 43, and SBCCI Standard Building Code Chapter 10.

Many fire-resistive assemblies use cellular floors, decks, and suspended ceilings. Openings for recessed lighting fixtures, electrical outlets, and air vents might be included in these fire-resistive assemblies. The ceiling usually is, but might not be, tested as part of the assembly. In any event, the details of the construction are important if required fire resistance is to be achieved. Fire-resistive assemblies tested by UL are "classified." This is the same approval as "listed" but is connotative of testing for a very specific application in a specific manner. It is essential that a listed fire-resistive assembly conform to the tested specimen in all particulars after field installation. The materials and dimensions must be at least equal to those of the test specimen and constructed per the classified design. Any holes or apertures in the fire protection must be limited in number, area, and distribution in the same manner as the test assembly. Engineering analysis also must show that fire resistance of an assembly is not critically affected by a proposed arrangement of openings for lighting fixtures or ducts. The code permits openings for outlets, air-conditioning ducts, and similar equipment where such openings do not exceed the percentage of the ceiling area established by fire test data. Because such openings allow only a small amount of heated gases generated by a fire to enter a concealed space, their hazard is considered negligible. Openings of this type are often described as a *membrane penetration*. Obviously, no such penetration shall extend through an entire wall or floor and ceiling assembly. The design and construction of fire-rated assemblies are quite specific, even including specific nails and screws. Minor changes in design, such as thickness or the method of fastening of construction materials, can have a significant effect on the assembly's behavior during a fire. Wires supporting a suspended grid ceiling, for example, are critical; for this reason, the *NEC®* does not allow such wires to be used for support of electrical wiring when it is a rated ceiling (see *NEC®* Section 300-11[a]). If a wall or ceiling is perforated for the passage of raceways or for work space around them, the fire resistance of the wall or ceiling is effectively destroyed unless the opening around the penetrating item is properly sealed. This danger is recognized in the *NEC®* Section 300-21, which states, "Electrical installations in hollow spaces, vertical shafts, and ventilation or air-handling ducts shall be so made that the possible spread of fire or products of combustion will not be substantially increased. Openings around electrical penetrations through fire-resistance rated walls, partitions, floors, or ceilings shall be firestopped using approved methods to maintain the fire-resistance rating." Firestopping must be installed in accordance with the local adopted building and fire codes.

Because most plumbing, air handling, electrical, communications, and other building services must pass through fire-resistant assemblies, coordination of building design, type of materials, and installation methods for these systems is essential to minimize the danger of spread of fire. Changes in design of or in materials used in a fire-rated assembly should not be introduced without preapproval of the AHJ.

Plans for proposed buildings might not indicate all eventual penetrations. The absence of detailed provisions for electrical or communications systems in the floor plan does not mean that such systems will not be installed later. If the electrical and communications circuits are to be added after construction is completed, it might be necessary either to construct chases (shafts) in fire-rated floors, walls, and ceilings for cables or raceways to pass to other areas or to construct openings in the walls, floors, or ceilings. Such penetrations through an entire assembly might void the intended fire protection of the construction and allow the spread of any of the products of combustion (smoke, hot or toxic gases, or flame). Creation of passageways through which smoke, gases, and heat can pass only increases the hazards to the building's occupants. It is the responsibility of the AHJ to determine whether a particular firestop method or material is to be approved. This is a significant responsibility for the AHJ, who usually must look to others for information to verify that a particular method or material has been adequately evaluated.

The building codes provide two recognized methods for sealing openings around penetrating items. These are through-penetration fire-stop systems and annular space filler materials. Through-penetration systems are available for both combustible and noncombustible penetrating items. Annular space fillers can be used only with noncombustible penetrants. Why the difference?

When combustibles such as nonmetallic-sheathed cables, insulated conductors in cable trays, and PVC conduit are exposed to heat, they burn or melt away, and materials surrounding them fall away, leaving a hole through which the fire could continue to travel either vertically or horizontally to adjoining floors or compartments. In the 1970s a disastrous cable fire spread through the Brown's Ferry Nuclear Plant in just this manner. Because of that event, new methods of sealing were developed; among them are materials that swell up when exposed to heat and fill the hole as the nonmetallic materials burn away. This swelling process is called *intumescing*. Other through-penetration systems are the endothermic type. This means that they release chemically bound molecules of water when exposed to heat. Both types must be tested in accordance with ASTM E815. UL Standard 1579 is the same as ASTM E815 ("Fire Tests of Through-Penetration Firestops"). Through-penetration systems classified by UL are found in the "Orange Book" (*Fire Resistance Directory*). This directory is very explicit about the assembly design, the size and type of the penetrating item(s), the size of the opening around the item, and the thickness and construction of the protection materials. It is very important that the listing criteria be followed precisely as the systems are designed to ensure enough volume to fill the opening when the penetrating item burns or melts away or to control the temperature on the unexposed side. Underwriters Laboratories is explicit that individual components are not rated and that they are not intended to be intermixed among systems. The classified rating applies only to the complete system. There are some who think that through-penetration fire-stop systems prevent the passage of smoke and toxic gases. Although this might be true in some cases, the E815 test does not provide for determining this, so it is an unknown.

Through-penetration fire-stop systems are absolutely necessary for combustible penetration. For noncombustible penetrations, such as EMT, IMC, and rigid conduit, either through-penetration systems or annular space fillers can be used. The systems for metallic penetrations are also found in the UL "Orange Book." To use this book, for both combustibles and noncombustibles, you will need to know the specifics of the assembly being penetrated, such as the fire rating, type of studs, thickness of gypsum or concrete, and the size and type of penetrating item(s).

The "Orange Book" indicates an "F" rating and a "T" rating. The "F" rating is based on the time period for which the system prohibits flame passage to the unexposed side (as evidenced by flame occurrence) in conjunction with an acceptable hose stream test performance.

Unless the penetration is contained in a wall cavity, you might also need to know the "T" rating given to the specific system. The "T" rating is the test time expired before the sealant and penetrating item reach a temperature of 325°F above ambient on the nonfire side of the assembly in addition to containing the flame and passing the hose stream test. Hose stream testing is the application of a specified water stream from a fire hose to evaluate the structural integrity of the assembly after fire exposure. It is not relative to firefighting activities during an actual fire. You should consult the applicable building code for exact language regarding "T" ratings as this is still in the process of change and varies in different codes.

Annular space fillers, which are permitted to be used only with noncombustible penetrants, do not require "T" ratings. However, the building codes do require that, to qualify as annular space fillers, materials be able to withstand ASTM E119 time-temperature conditions under positive pressure for the full period for which an assembly has been rated without igniting cotton waste placed in contact with it.

Common construction materials have long been used with great success as annular space fillers. These are products such as cement, mortar, grout, handyman caulk, mineral wool, and even joint compound. As an aid to industry, the "NEMA Steel Rigid Conduit and EMT Section (5RN)" sponsored an investigation at Underwriters Laboratories to document the performance of these longtime annular space fillers.

These were generic-type tests and easily affirmed the performance of these materials. On the basis of this testing, the three national model building codes have added language that automatically accepts cement, mortar, and grout for sealing openings in masonry assemblies. The applicable code should be reviewed for specifics because there are some variables in the precise language. A summary (or full) report on this annular space testing is available from NEMA, Section 5RN, 1300 N. 17th Street, Suite 1847, Rosslyn, VA 22209. Ask for "UL Special Services Investigation Fire NC 546 Project 90NK11650."

This report documents that everyday joint compound can maintain the fire-resistance rating of a gypsum wallboard assembly for two hours. The report shows that even with the damage to the gypsum board (½-inch sheetrock) on the fire side of the test specimen that occurred when the furnace exploded, the fire-resistance rating of two hours was achieved. It is a good document for use in securing AJH approval when planning to seal with annular space filler.

Electricians and designers must become educated on the sealing of openings around electrical wiring in order to comply with *NEC®* Article 300-21 and with the building codes.

Combustibility of Raceway Materials and Conductor Insulation

In the last twelve to fifteen years, a great deal of attention has been given to hazards of fire relative to combustible electrical systems (along with combustible construction furnishings and finishing materials). Much of this attention was created by the expansion in the *NEC®* of nonmetallic wiring methods within buildings. The degree of the hazard remains controversial, as does the risk involved. Certainly, individuals draw different conclusions from the same set of facts; additionally, risk tolerance varies from person to person. Otherwise, we would not have deep-sea divers and fighters of oil well fires. Other questions have also arisen in this controversy, such as how many lives is it acceptable to lose—we cannot feasibly get to zero. The age-old response then becomes, your family or mine? In this section we look at some of the issues involved. The user, designer, or jurisdiction then must determine the acceptable level of risk for the particular situation.

The *NEC®* is a minimum code and allows choices in wiring methods. However, in a few cases it has directly addressed the matter of fire concerns. One specific area is in plenums and other environmental airspaces. The whole construction code arena recognizes the dangers inherent in allowing the products of combustion (smoke, heat, and toxic gases) to spread throughout a building by way of the air handling system. Materials allowed in such spaces are carefully controlled (although a series of events in recent years is making this control borderline).

Article 300-22 of the *NEC®* controls the electrical system in environmental airspaces. The wiring methods in ducts or plenums used for environmental air (Section 300-22[b]) must be EMT, IMC, rigid metal conduit, flexible metallic tubing, MI cable, or type MC cable employing a smooth or corrugated impervious metal sheath without an overall nonmetallic covering. To connect physically adjustable equipment and devices permitted in these areas, lengths of flexible metal conduit or liquidtight flexible metal conduit not over 4 feet are permitted.

A plenum is defined in the *NEC®* as a "compartment or chamber to which one or more air ducts are connected and which forms part of the air distribution system." The BOCA National Building Code contains this definition: "An enclosed portion of the building structure which forms part of an air distribution system and is designed to allow the movement of air."

Thus, it is important to recognize that a ceiling space through which the building air moves is considered a plenum. If there is ever any doubt relative to the *NEC®* definition, remember that the adopted building code governs.

In environmental airspaces, other than ducts or plenums (Section 300-22[c]), other wiring methods are added to those named above. These include totally enclosed nonventilated, insulated busways having no provisions for plug-in connections, type AC cable, and other factory-assembled multiconductor control or power cable that is specifically listed for the use. Some of these are called "plenum cable," which is a misnomer because they are not for use in actual plenums. One other means of enclosing the conductors, beyond those conduits named in Section 300-22(b), is surface metal raceway or wireway with metal covers or solid bottom metal cable tray with solid metal covers. For minor exceptions, refer to Section 300-22(c).

What are some of the concerns regarding combustible wiring methods, whether or not in environmental airspaces? The primary nonmetallic materials used in buildings for raceways and conductor insulation are PVC and nylon. When PVC reaches a temperature in the range of 450°F to 500°F, it starts to chemically break down (thermally decompose), although some formulations might not ignite until almost 1,000°F. This decomposition releases a very corrosive and irritating hydrogen chloride gas (HCL) and at this point might not even produce visible smoke. This HCL is very irritating to the eyes, nose, and lungs of those who are exposed to it. As the decomposition progresses, thick black smoke is also produced; actual burning (flaming) is not required. Between the smoke obscuring the path of escape, the burning eyes, and choking sensation, it becomes very difficult to evacuate the area. Even if one is able to get out, sufficient exposure to HCL as a product of combustion (attached to soot particles) can produce severe respiratory problems that are frequently irreversible and at times lead to delayed death. (In toxicity tests, many animal deaths from PVC combustion do not occur until several days later.) PVC also produces large amounts of carbon monoxide. Although nonplasticized PVC, such as that used in raceways, is ignition resistant, it will burn once the chlorine and other fire retardants are driven off. Plasticized PVC, used in conductor insulation, burns more quickly. Another produced gas that is of concern is benzene, which is both flammable and carcinogenic. Burning nylon produces cyanide.

An additional effect of HCL is its corrosive effect on electrical contacts and other electrical and electronic equipment. Deposits left by burning PVC in equipment rooms, even after thorough scrubbing, have been known to cause corrosion weeks and months later. Computer equipment exposed to PVC combustion products during testing has been damaged. The corrosive effects on electrical contact of HCL from overheated conductors could affect operation of safety and signaling equipment. Testing on the nonmetallic side of the issue shows that HCL from PVC adheres to walls, floors, and room objects. The question is how much of the HCL is therefore removed from the atmosphere—how much under the great variety of fire conditions is the unanswered question. Certain *NEC®* articles also require that nonmetallic conduit be placed behind a thermal barrier to inhibit its exposure to heat for a projected 15 minutes. In such cases, it must be ensured that inadvertent openings are not made and left unsealed or that thermal barrier ceiling tiles are not removed.

Noncombustible (metallic) raceways have long been the recognized and proven method of inhibiting fire spread by way of the electrical system. Even if the conductor insulation inside burns, the spread will be limited, the smoke will be more contained, and the raceway itself will not contribute smoke or hazardous gases.

It is incumbent on users to evaluate the specific set of conditions for an electrical system, including the number and type (as related to their ability to exit in an emergency) of occupants that might be anticipated in a building and its design and thus to determine the level of risk they want to take relative to combustible materials. It is important to remember that state supreme courts have found that compliance with adopted codes alone does not necessarily preclude fire-injury liability. This does not apply only to the electrical system—it is a consideration for all design.

Wiring in Ducts, Plenums, and Other Air Handling Spaces

The provisions for installation of electrical wiring and equipment in ducts, plenums, and other air handling spaces are generally covered in *NEC®* Article 300, Section 300-22. It should be noted, however, that often building codes and local fire codes take precedence and are usually more restrictive. The *NEC®* Article 100 defines a plenum as "a compartment or chamber to which one or more air ducts are connected and which forms part of the air distribution system."

It is advised that the designer or installer check with the chief building official for guidance and direction before encroaching on these areas. No wiring of systems of any type shall be installed in ducts used to transport dust, loose stock, or flammable vapors. The *NEC*® Section 300-22(a) restricts wiring of any type from being installed in any duct or shaft containing only such ducts used for vapor removal or for ventilation of commerical-type cooking equipment. Wiring methods approved for ducts or plenums used for environmental air are type MI cable, type MC cable employing a smooth or corrugated impervious metal sheath, electrical metallic tubing, flexible metal conduit, intermediate metal conduit, or rigid metal conduit and are permitted to be installed in ducts or plenums specifically fabricated to transport environmental air. Flexible metal conduit and liquidtype flexible metal conduit not to exceed 4 feet is permitted where adjustable equipment and devices are located in these chambers. The connectors used for flexible metal conduits shall close openings in the connection. Equipment and devices shall be permitted within such ducts and plenum chambers only where necessary for the direct action or sensing of the contained air. Where such equipment or devices are installed and illumination is necessary to facilitate maintenance and repair, only enclosed gasket-type fixtures are permitted. In accordance with Section 300-22(c), the other space used for environmental air, which is used for purposes other than ducts and plenums, is described in Section 300-22(b). An example is the space over a suspended ceiling that is used for environmental air handling purposes. The wiring methods approved in these areas are totally enclosed nonventilated insulated busways having no provision for plug-in connections, wiring methods consisting of MI cable, type MC cable without an overall nonmetallic covering, type AC cable, or other factory assembly multiconductor control or power cable that is specifically listed for the use. Other type cables and conductors shall be installed in electrical metallic tubing, flexible metallic tubing, intermediate metal conduit, rigid metal conduit, flexible metal conduit, or, where accessible, surface metal raceway or wireway with metal covers or solid bottom metal cable tray with solid metal covers. Electrical equipment with a metal enclosure or with a nonmetallic enclosure listed for use and having adequate fire resistance and low smoke-producing characteristics and associated wiring materials suitable for the ambient temperature shall be permitted in such areas, unless specifically prohibited in other sections in the *NEC*®. There are limited exceptions to these rules. Data processing raised floor areas are covered in *NEC*® Article 645, which requires that all the conditions in Section 645-2 be met before the area can be classified as a data processing room. The requirements as noted in *NEC*® Section 645-2 are that disconnecting means must be installed to disconnect all electronic equipment in the room and all power to dedicated HVAC heating, ventilating, and air-conditioning equipment serving the room and to cause all required fire and smoke dampers to close. The disconnects must be grouped, identified, and controlled from locations readily accessible at all principal exit doors. In addition, if a separate heating, ventilating, and air-conditioning (HVAC) system is provided that is dedicated to the electronic computer and data processing equipment's use and is separated from other areas of occupancy, except other HVAC equipment, it can serve these areas, provided that fire/smoke dampers are installed at the point of penetration of the room boundary. Such dampers must operate on the activation of smoke detectors; also a disconnecting means is required. Only listed electronic computer and data processing equipment is installed in the room. The room is occupied only by those personnel needed for maintenance and the function operation of that room, and the room is separated from other occupancies within that building or floor by fire-resistant rated walls, floors, and ceilings with protected openings. The building construction shall comply with the applicable building codes. Good engineering design and close coordination with the AHJ responsible for these occupancies, such as the chief building official, fire marshal, and the chief electrical inspector, should all be consulted before applying Article 645 to an electrical system. Remember that *NEC®* Article 645 permits a more lenient wiring method; therefore, greater caution must be taken to confine a fire to that area and to ensure that it does not propagate (spread) to the rest of the building. Steps should be taken to ensure easy egress for those working in that room in the event of an emergency.

Raceway Allowable Conductor Fill

The allowable conductor fill requirements are found in Chapter 9, the tables (including the notes to those tables), and Appendix C of the *NEC®*. With all types of conductors except lead covered, three conductors or more in a raceway is limited to 40% of the cross-section of the conduit or tubing. Other allowable percentages of fill are listed in Table 1 of Chapter 9. Equipment grounding or bonding conductors when installed must be included in the calculations. The actual dimensions shall be used in making the calculation, whether these conductors are insulated or bare. Conduit nipples not exceeding 24 inches are permitted to be filled to 60% of the total cross-sectional area. Conductors not included in Chapter 9 shall be based on their actual dimensions for the purpose of making the calculations. Conductors' dimensions are found in *NEC®* Table 8 of Chapter 9. Multiconductor cable of two or more conductors must be treated as a single conductor when calculating the percentage of conduit fill. For cables that have an elliptical cross-section, the calculations shall be based using the major diameter of the ellipse as the circle diameter. The trade size of the raceway in *NEC®* Table 4 of Chapter 9 shall be used as the internal diameter in inches.

Example

As an example, what size raceway is needed to enclose nine number 12 THW conductors in an intermediate metal conduit? Answer: ½ inch. By contrast, electrical nonmetallic tubing would require a ¾-inch raceway. (Reference: *NEC®* Chapter 9, Table 1 and Appendix C, Table C2.) As another example, what size raceway is needed to enclose four 500 kcmil THWN and one 300 bare conductor? Answer: The 1996 *NEC®* has revised all the tables in Chapter 9. As a result, the type of raceway must be known. For example, for EMT the minimum size would be 3 inches. By contrast, for a Schedule 80 PVC conduit the minimum size would be 3½ inches. (Reference: *NEC®* Chapter 9, Tables 1, 4, 5, and 8.)

It should be noted that although the raceway is sized according to the allowable conductor fill in Chapter 9 of the *NEC®*, there are other considerations necessary in sizing the conductors before the raceway system itself is sized, such as the ampacity adjustment factors in Article 310 for more than three conductors per raceway. There are ambient temperature considerations that require ampacity correction where the ambient temperatures exceed levels as listed at the bottom of each of the ampacity tables as they apply. Good engineering practice, however, recommends that when all these things are considered, the raceway system should be oversized by 50% to handle future expansion and growth to the electrical system.

Raceway System

About 100 years ago, as we entered the 1900s, our great electrical industry was growing like wildfire. This was a result of the great inventors such as Thomas Edison, holder of hundreds of patents promoting DC voltage; Nikola Tesla, inventor of the three-phase motor and a promoter of AC voltage; and William Merrill, founder of Underwriters Laboratories, along with George Westinghouse, and others (Figure 14–17). As a result of this rapid growth in the use of electricity, there were many fires, electrocutions, and electrical accidents almost

Figure 14–17 Early inventors searched for the optimum raceway to carry electrical conductors. They found tubular steel conduit to be that raceway. Rigid steel conduit has provided that need for more than 100 years.

everywhere. With arcing and sparking and resultant fires occurring almost daily and with keen competition between Edison (DC) and Tesla (AC), a group got together in 1881 to discuss the safer use of electricity. The National Association of Fire Engineers met in Richmond, Virginia, and from this meeting came a proposal that served as a basis for the first *NEC*®. In the earliest beginnings of the electrical era, before the turn of the twentieth century, men began to design various means of distributing the wonderful servant made practical by the work of these great inventors—electricity. Ordinary electrical conductors strung without protection soon proved faulty, dangerous, and wasteful. Some means by which they could be enclosed and protected permanently against mechanical and electrical damage was clearly needed. The black-painted steel pipe that originally ran through buildings to provide gas for lighting was already in place and provided the first protection for electrical conductors. From that evolved today's galvanized rigid steel conduit and later the lighter-weight steel electrical metallic tubing as wiring became more widespread. The earlier conduits were of zinc with copper elbows; then spiral-wrapped paper tubing with brass joining sleeves was introduced. As the shortcomings of these conduits were revealed, brass-enclosed paper tubing with brass elbows and couplings was introduced. But this, too, was found to be unsatisfactory, mainly because of inadequate physical protection and mechanical strength. The concept of a conduit into which electrical conductors could be pulled in and out remained as worthwhile, but the sole question to be answered was, What type of material and processing would provide the best and most economical conduit?

Iron-armored conduit was introduced in 1894. This consisted of 10-foot lengths of standard-weight wrought-iron gas pipe, with paper tube lining and threaded couplings and nipples. This conduit proved to be a great advance but was found to be still far from satisfactory in every respect. The linings reduced the usable interior area to an unsatisfactory

Figure 14–18 Raceway systems installed today are similar to those installed over fifty years ago. Many old systems are still being used today and continue to provide excellent physical protection. (Courtesy of Allied Tube & Conduit.)

degree. Linings of paper or fiber would crumble, and wooden linings would splinter, making field bending impossible. These linings were not moisture-proof and were expensive.

The introduction of good rubber-insulated electrical conductors in 1897 made possible the development of unlined steel conduit. Here at last was a conduit that provided full mechanical and electrical protection for conductors. The system could be readily expanded and furnished maximum design flexibility, as well as a host of other desirable features. Conduit also serves as an excellent equipment-grounding conductor. The electrical system could be changed by pulling in new conductors without damaging the walls, floors, and ceilings. These desirable features in one easily installed, convenient raceway system led to the widespread use of steel conduit. Although the quick acceptance of this new steel conduit by the electrical industry was followed by rapid improvements in the product, it was not without its problems. The gas fitters immediately claimed the right to install their wrought-iron gas pipes, and there ensued great craft disputes between the electrical craftsmen and the gasfitter craftsmen. This was finally resolved, and the electrician was allowed to install conduit. About this same time, groups began to cut wrought-iron gas pipe into 10-foot lengths and bootleg it over the counters to the electricians as conduit. The problem was that electrical conduit was rougher on the exterior and smoother on the interior, whereas wrought-iron gas pipe was just the opposite. The conduit interior was important so as not to tear the rubber insulation on the conductors as they were being pulled in. Fortunately, with the rapid advance in the industry and improvements in the product, these problems were soon far behind.

The first exterior coating to prevent oxidation was enamel. In 1902 the first electro-galvanized rigid steel conduit was introduced. In 1908 sherardized steel conduit appeared, followed by hot-dip galvanizing in 1912. A little later, the metallizing process was introduced.

Advantages of Raceway Systems

Since 1905 the two primary functions of electrical raceway have been to facilitate the insertion and extraction of the conductors and to protect the conductors from mechanical injury (see Figure 14–18).

Today's raceway systems combine these functions with the need for a high degree of safety and greater distribution and utilization flexibility. It is the responsibility of the designer and installer to select the correct raceway for each application, and all things must be considered. *NEC®* Section 110-3 requires that during the judging of equipment, considerations such as the following shall be evaluated: The suitability of the installation must be considered and the use and conformity of equipment applicable with the provisions of the code; the suitability of the equipment for the specific purpose, environment, and application of use; and the suitability of the equipment as evidenced by listing or labeling. The mechanical strength and durability, including parts designed to enclose and protect the other equipment, is evaluated for suitability. The equipment is evaluated for adequate wire bending and connection space, electrical insulation, and heating effects under normal use and abnormal use. Equipment is evaluated for arcing effects. It is classified by type, size, voltage, current capacity, and specific use. Other factors include the practical safeguarding of persons likely to come in contact with the equipment. *NEC®* Section 110-3(b) states that listed and labeled equipment shall be used or installed in accordance with any instructions included in the listing or labeling. Therefore, improper selection of a raceway system or improper installation of that system is an *NEC®* violation, and the product might not provide the service or safety for which it is intended. Selection of a wiring method should include consideration of the type and design of the building and the number and capabilities of the occupants relative to escape in the event of a fire. Combustibility aspects of a raceway or other wiring method become important because the wiring system is installed throughout the building (see Figure 14–19). Also, in some installations there are large quantities of wiring overhead. Often many branch circuits and feeders are routed from the equipment room over or adjacent to corridors.

Types of Metallic Wiring Methods Systems

Rigid Metal Conduit (*NEC®* Article 346)

NEC® Article 346 covers rigid metal conduit, also called galvanized rigid conduit (GRC), rigid metal conduit (RMC), and aluminum rigid metal conduit (ARC) (see Figure 14–20). The most common is the heaviest weight, tubular steel conduit (GRC). It is smooth walled and galvanized on both the inside (ID) and the outside (OD). The galvanizing can be applied by the hot-dip process (dipping into a pot of molten zinc) or the electro-galvanizing process (electro-deposited zinc). Galvanizing provides protection against corrosion. The degree of corrosion resistance is influenced by environmental conditions, such as moisture and chemical exposure. Also covered under this article is nonferrous metal conduit. Both products are evaluated to the UL 6 Standard. The nonferrous product most commonly produced is aluminum rigid metal conduit and silicone bronze conduit and is evaluated to ANSI C80.5.

UL 6 and ANSI C80.1 are the standards for GRC and RMC. The international standard is IEC 981. These IEC soft metric conversions are the metric designations ½ = 16, ¾ = 21, 1 = 27, 1¼ = 35, 1½ = 41, 2 = 53, 2½ = 63, 3 = 78, 3½ = 78, 4 = 103, 5 = 129, and 6 = 155. These conduit types are currently listed and produced to the UL and ANSI standards. Both have been adopted by the federal government. The federal specification was formerly WW–C–581, Class 1/Type A. The dimensions and weights are shown in Table A–10 of the Appendix.

Galvanized rigid steel conduit is a threaded product, with threads conforming to ANSI B2.1. and *NEC®* Section 346-7(b). The UL listing requires that a listed piece of conduit have a coupling attached to one end. This coupling is electro-galvanized to avoid zinc buildup on the interior threads that might occur with hot-dipping. All rigid conduit is

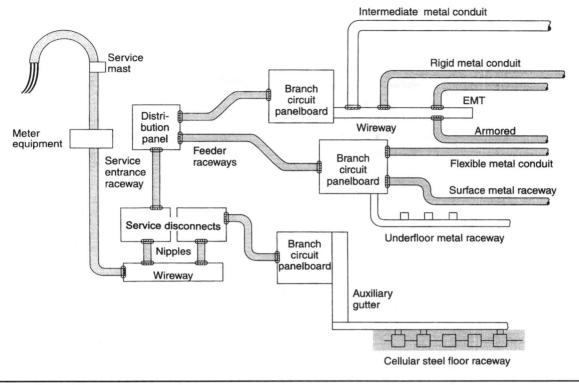

Figure 14–19   Wiring methods for 600 volts or less per *NEC®* Chapter 3 and for over 600 volts as per Article 710.

Figure 14–20   Rigid steel conduit. (Courtesy of Allied Tube & Conduit.)

shipped with the coupling on one end and the open threads on the opposite end protected by color-coded plastic caps. These caps are color-coded to provide easier field identification. The red caps indicate quarter sizes, such as ¾ inch, 1¼ inches, and so on. The black caps are for ½-inch sizes such as ½ inch, 1½ inches, and so on. The blue caps are for the even sizes, such as 1, 2, 3, 4, 5, and 6 inches.

GRC is produced in trade sizes ½, ¾, 1, 1¼, 1½, 2, 2½, 3, 3½, 4, 5, and 6 inches. It should be noted that these trade designations are *not* actual OD or ID dimensions. The ID for ½-inch GRC, for example, is 0.632 inches; the OD for ½-inch GRC is 0.840 inches. Neither dimension is the .500 one might expect. The actual diameters of all the recognized raceways can be found in *NEC*® Chapter 9, Table 4.

Article 346, and the *NEC*® in general, recognize the use of rigid steel conduit in *all* applications, including hazardous (classified) locations. The use of galvanized rigid steel conduit is only exempted where severe corrosive conditions exist. Even then, additional supplementary coatings, such as PVC-coated rigid metal conduit, can be used for those applications. Rigid steel conduit is the only conduit that can be used in all locations in accordance with the *NEC*® without exceptions. Steel rigid conduit and electrical metallic tubing are the most widely used raceway products in production today. It is common to find GRC conduit still in use in buildings built around the turn of the twentieth century and still providing good, safe service for the installation. Many buildings being remodeled today utilize the existing rigid metal conduit systems in those buildings for all or a portion of the new wiring being installed.

Galvanized steel rigid conduit evolved from black gas pipe, which was already in place and a logical means of protecting those "newfangled" electrical conductors Tom Edison was touting! GRC is considered to be the premier electrical wiring method. It provides superior physical protection for conductors, is not affected by hot or cold temperatures (there are no code or standards limitations), and can be indoors, outdoors, and underground, concealed or exposed.

Rigid metal conduit is a code-recognized, equipment-grounding conductor in accordance with *NEC*® Section 250-91(b).

Intermediate Metal Conduit (IMC)

Intermediate metal conduit, commonly know as IMC, is covered in *NEC*® Article 345. It was developed in the early 1970s by Allied Tube and Conduit. It originally was intended to be an entirely separate system but has evolved with threads and dimensions that make it interchangeable with GRC (see Figure 14–21). In fact, the same couplings are used on both IMC and GRC. The *NEC*® Articles 345 and 346 are virtually identical, and the *NEC*® recognizes both products for exactly the same applications, without limitations, including the metric designators; however, IMC is only manufactured in sizes ½ (16) through 4 (103).

The standards for listing and production of IMC are UL 1242 and ANSI C80.6. UL 6 has been adopted as a federal specification. IMC was formerly covered by WW-C-581 as Class 2/Type A (ANSI C80.6 has not yet completed the adoption process).

IMC is galvanized on the OD and has a UL-approved corrosion resistant organic or inorganic coating on the ID. Although the IMC wall is less than the wall thickness of GRC, its method of manufacture and steel chemistry provide physical strength equivalent to, or often greater than, that of GRC, thus providing equivalent physical protection. Listed IMC is also a conduit threaded to ANSI B2.1 with an attached coupling. The color coding for the protective end caps is ¼-inch-sizes green, ½-inch-sizes yellow, and even-sizes

Figure 14–21   IMC intermediate steel conduit. (Courtesy of Allied Tube & Conduit.)

orange. This provides easy visual differentiation between IMC and GRC. Further identification is provided by marking the letters "IMC" on the pipe at regular intervals. The ID of IMC is slightly larger than that of GRC, with ½-inch having a normal .660 ID; the OD normal for ½-inch is 0.815 inches, making wire pulling in IMC easier. The dimensions for nominal trade sizes of IMC can be found in Figure A–10 of the Appendix.

Wall thickness comparison is 0.070 inches for ½-inch IMC and 0.104 inches for ½-inch GRC. Compatible threads are possible because less steel is cut away in threading IMC. As with rigid conduit, each length is designed to be 10 feet, including the coupling.

IMC has become a very popular replacement for GRC in that it provides the equivalent physical protection, has a larger ID, and can be installed more easily because of its lighter weight. It is a totally different product than GRC and requires adjustment to common standard threading machines before threading and special bending equipment or adapters on standard bending tools. However, once a contractor is set up to install IMC, it usually is found to be easy to install.

Electrical Metallic Tubing (Article 348)

Electrical metallic tubing, commonly known as EMT, is the lightest weight tubular raceway manufactured. It is an unthreaded plain end product that is joined together by the use of set-screw, indentation, or compression-type connectors and couplings. EMT is produced as galvanized steel or aluminum. The coupling is not provided as a part of the listed EMT. EMT was developed before World War II but did not gain popularity until the start of the war, when the entire nation was trying to conserve steel wherever possible. Soon after the use of EMT began and the tradesmen learned how to bend and install the product, it gained the massive popularity it enjoys today (see Figure 14–22). It is widely used throughout the industry for branch circuit and feeder raceways. It is very versatile in that it can be altered, reused, and redirected with ease because of the unthreaded design. The conductors can be inserted and extracted easily because of the very smooth interior, making conductor installation fast and easy. The internal diameter is sufficiently large to provide ease in pulling the number of conductors permitted in the *NEC*®. EMT has one of the larger IDs among the many raceway types being manufactured today. Despite the lightweight steel construction, it provides substantial physical protection and can be used in most exposed locations, except where subject to *severe* physical damage.

EMT is produced in sizes ½ through 4. The metric designators are ½ = 16, ¾ = 21, 1 = 27, 1¼ = 35, 1½ = 41, 2 = 53, 2½ = 63, 3 = 78, 3½ = 91, and 4 = 103.

The listing and production standards for EMT are UL 797 and ANSI C80.3. These standards have also been adopted by the federal government, replacing WW-C-563.

Electrical metallic tubing has a galvanized OD and a UL-approved corrosion-resistant organic or inorganic coating on the ID. For identification purposes, each 10-foot length is marked "EMT."

Figure 14–22 EMT electrical metallic tubing. (Courtesy of Allied Tube & Conduit.)

Again, the dimensions are trade size designations only, not actual ½-inch. ID is .620 inch, and OD is a nominal 0.706 inch. Other weights and dimensions for electrical metallic tubing can be found in Figure A–10 of the Appendix.

The *NEC*® recognizes EMT for both exposed and concealed work. It can be used in most applications, expect where subject to severe physical damage or where subject to cinder concrete or cinder fill subject to permanent moisture, unless protected on all sides by a layer of noncinder concrete at least 2 inches thick.

It is also limited in some hazardous locations. EMT is permitted to be installed in wet locations when used with proper fittings. Both rain-tight and concrete-tight fittings are available and must be used in wet locations and in poured concrete.

The support requirements for electrical metallic tubing can be found in *NEC*® Section 348-12. Bending shall be in accordance with Section 346-10 and must not exceed four quarter bends or 360° total between pull points. In most cases, EMT must be supported at least every 10 feet and within 3 feet of outlet boxes, junction boxes, and conduit bodies. In ceiling joists, bar joists, and other areas where structural members do not readily permit support within 3 feet, a distance of 5 feet from each outlet box, junction box, or conduit body is permitted.

Special Feature Tubular Steel Conduits

Integral Coupling Steel Raceways

IMC and GRC can be purchased with an integral coupling, such as Kwik-Couple, which allows joining of two lengths of rigid or IMC together by turning the coupling rather than the conduit. This coupling performs similar to that of a union (Erickson) in that it can be tightened without rotating the conduit (see Figures 14–23 and 14–24). This type conduit can provide substantial savings where routing would cause difficulty in joining the conduit lengths together or where several unions would be required because of the architectural

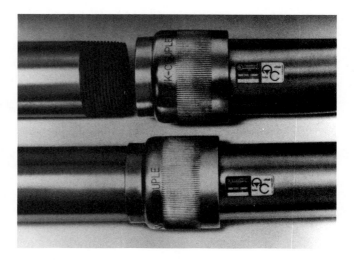

Figure 14–23 Integral coupling steel rigid and IMC conduit meets Sections 250-91(b) for grounding requirements of the *NEC*®. (Courtesy of Allied Tube & Conduit and Triangle Wire and Cable Company.)

Galvanized Electrical Metallic Tubing (EMT)

Industry Standards:
UL 514B - Fittings for conduit and outlet boxes
UL 797 - Electrical metallic tubing
Federal Spec. WW-C-536-A - Conduit, metal, rigid
      electrical, thin-wall steel type
ANSI standard C80-3 - Electrical, metallic tubing
      zinc coated

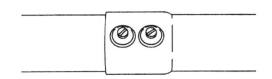

Galvanized Electrical Metallic Tubing (EMT) Compression Uni-Couple

Industry Standards:
UL 514B - Fittings for conduit and outlet boxes
UL 797 - Electrical metallic tubing
Federal Spec. WW-C-536-A - Conduit, metal, rigid
      electrical, thin-wall steel type
ANSI standard C80-3 - Electrical, metallic tubing
      zinc coated

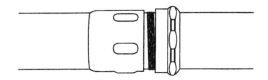

Figure 14–24 An example of two different types of integral coupling EMT electrical metallic tubing. (Courtesy of Allied Tube & Conduit and Triangle Wire and Cable Company.)

design or obstructions in existing installations. Many installers integrate these special feature raceways with the traditional raceways to accomplish an easier and better installation. Where field bends or architectural appurtenances have made joining difficult or impossible without a union (Erickson) in the past, these new state-of-the-art products make installations easy and fast.

EMT is available with a belled-end fitting with installed set screws or integral compression coupling, which eliminates the need for a separate coupling. Substantial savings can be achieved in installed cost.

Coated Rigid Conduit and Intermediate Metal Conduit (IMC)

The coated metallic raceways that are manufactured today are manufactured in accordance with NEMA Standard RN-1 for "Polyvinyl Chloride (PVC) Externally Coated Galvanized Rigid Steel Conduit and Intermediate Metal Conduit." Coated metallic conduit must be installed in accordance with the article that covers the metallic conduit material used. For example, galvanized rigid metal conduit must be installed as per *NEC*® Article 346 (see Figure 14–25). Intermediate metal conduit must be installed in accordance with *NEC*® Article 345. A copy of NEMA Standard RN-1 is available through the National Electrical Manufacturers Association, Rosslyn, VA. This product adopts in whole or in part ANSI C80.1 and C80.6. It also adopts ASTM D159/87 test methods for dielectric breakdown voltage and dielectric strain of solid electrical insulating materials at commercial power frequencies, D638-87, the test method for tensile properties of plastics; D1790-83, the test method for brittleness temperature of plastic film by impact; D2240-86, the test method

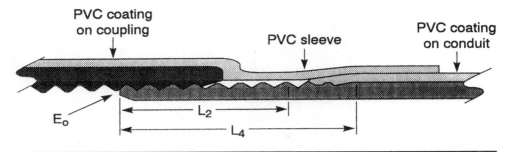

Figure 14–25 PVC-coated rigid conduit is ideal for high corrosive areas; areas subject to physical damage can be coated both inside and out. (Courtesy of Robroy Industries.)

for rubber property durometer hardness; G6-83, the test method for abrasion resistance of five-ply coatings; G10-83, the test method for specific bendability of pipeline coatings; G23-81, the practice for operating light and water exposure apparatus (carbon-arc type) for exposure of nonmetallic materials; Underwriters Laboratories UL6; and UL1242.

PVC-coated rigid conduit was first introduced in 1961 by Robroy Industries. Today there are three major manufacturers of PVC-coated steel conduit Robroy Industries, Occidental Coating Company, and Perma-Cote Industries. These major manufacturers of coated conduit have recognized the need for a wiring method that provides complete corrosion protection and excellent installation qualities. PVC-coated conduit combines the strength of metal with the corrosion-resistant qualities of a bonded plastic coating for a permanent, trouble-free installation. It is an attractive product and identifies your electrical conduit system. In service it continues to create economies through complete and lasting corrosion protection. The adhesive qualities of the PVC coating effectively prevent corrosive fume seepage—there is no undercreep or corrosion travel. Overlapping pressure sealing sleeves on couplings and condulets create tight pressure-sealed joints. You install it and forget it. There is no maintenance. The manufacturer begins by closely working with the designer and the installer to provide a complete material list. The coating process begins with listed rigid metal or intermediate metal conduit. Lighting fixtures can also be coated. Conduit bodies of all types and sizes are coated with overlapping pressure seals. Support channel, all hardware, such as straps, supports, are coated and corrosion-proofed the same as the conduit. Motor-starting switches, unions, and so on can be coated to provide a complete corrosion-resistant system. In addition, available from the manufacturer is a coating touch-up kit that consists of spray-on or paint-on protection that effectively repairs all damage to the original integrity and might even be of a heavier consistency than the original application (see Figure 14–26).

This product has been widely accepted for industrial facilities, both above-ground exposed and for direct burial in all environments. There is no equal for this state-of-the-art raceway system. Coated conduit offers the combination of excellent physical protection by using UL-listed galvanized rigid steel conduit or galvanized intermediate steel conduit plus unequaled corrosion protection by the addition of a 40-mil-thick PVC exterior coating. For severe environments in which corrosive vapors might also attack the conduit from the inside, a 2-mil-thick urethane interior coating can also be provided. These and other coatings can also be applied to rigid aluminum conduit to fit the requirements of each unique installation. **Note:** See Appendix for chemical resistance charts.

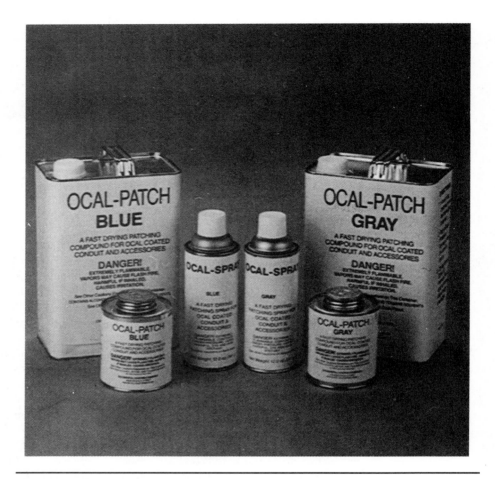

Figure 14–26 If PVC-coated conduit is damaged during installation, scrapes and cuts are easily patched with touch-up compounds available in spray cans, dabber, and brush-on. (Courtesy of Occidental Coating Company.)

Liquidtight Flexible Metal Conduit

NEC® Article 351(a) and UL 360 liquidtight flexible metal conduit is manufactured in trade sizes ⅜ (12) inch through 4 (103) inches for conductors in circuits of 600 volts nominal or less. liquidtight flexible metal conduit, which is suitable for direct burial and poured in concrete, is marked "Direct Burial." It is a raceway of circular cross-section having an outer liquidtight nonmetallic sunlight-resistant jacket over an inner flexible metal core with associated couplings, connectors, and fittings and approved for the installation of electrical conductors. Installations of liquidtight flexible metal conduit must comply with *NEC®* Articles 300 and 351 and specific sections where applicable of the *NEC®*. The marking requirements for this product must comply with *NEC®* Section 110-21. It can be used where listed and marked for the purpose for direct burial in earth and exposed and concealed work where the conditions of operation and maintenance require flexibility or protection from liquids, vapors, or solids. It shall not be used where subject to physical dam-

Figure 14–27   Underground nonmetallic duct bank installation. (Courtesy of Carlon, a Lampson Sessions Company.)

age or where any combination of ambient and/or conductor temperature will produce an operating temperature in excess of that for which the material is approved. Where it is installed as a fixed raceway, it shall be secured at intervals not exceeding 4½ feet and within 12 inches on each side of every outlet, box, junction box, cabinet, or fitting, except where permitted to be fished or not exceeding 3 feet in length for flexibility, and not exceeding 6 feet in length for fixture tap conductors. It is permitted as a grounding conductor where both the conduit and the fittings are approved for grounding. Where a bonding jumper is required, it must be installed in accordance with *NEC®* Article 250. Where liquidtight metal conduit is used for flexibility, an equipment-grounding conductor must be installed. It is permitted only as a grounding means in 1¼-inch trade sizes and smaller if the total length of the liquidtight flexible metal conduit and any ground return path is 6 feet or less and the conduit is terminated in fittings listed for grounding and the circuit conductors contained therein are protected by an overcurrent device rated at 20 amperes or less for ⅜-inch and ½-inch trade sizes and 60 amperes or less for ¾-inch through 1¼-inch trade sizes. Where used as a fixed wiring method, there shall not be more than the equivalent of four quarter bends, 360° total, between pull points (e.g., conduit bodies and boxes). Angle conductors are not permitted for use in concealed raceway installations.

Rigid Nonmetallic Conduit Schedule 40 and Schedule 80

Rigid nonmetallic conduit (PVC) (Schedule 40 and Schedule 80), intended for installation in accordance with *NEC®* Article 347 and UL 651, was originally introduced as an underground duct for use by electrical utilities in the late 1950s. At that time the duct was produced from high-impact polystyrene, which had excellent physical properties and corrosion resistance (see Figure 14–27). However, the styrene material did not have good fire-resistant properties, and by the early 1960s polyvinyl chloride (PVC) became the preferred material. Underwriters Laboratories first listed PVC Schedule 40 rigid nonmetallic conduit in 1962. The *NEC*® first recognized its use only for underground installations in 1968. In 1971 the *NEC*® expanded approved uses to include above-ground and in-building applications. Article 347, as it appears in the 1996 *NEC*®, remains virtually unchanged from the original 1971 edition (see Figure 14–28).

Rigid nonmetallic conduit, fittings, and accessories, also referred to as RNMC, are manufactured to NEMA TC-2 federal specifications WC1094A and UL 651 specifications and carry respective UL listings and UL labels. RNMC is nonconductive, sunlight resistant, and UL listed for exposed or outdoor usage (the use of expansion fittings allows the system to expand and contract with temperature variations), and it will not rust or corrode. The most commonly used RNMC is Schedule 40 rigid heavy wall PVC for underground applications encased in concrete or direct burial, exposed and concealed, rated for use with 90°C conductors and is manufactured in nominal sizes ½ inch through 6 inches. It is produced in 10-foot lengths but can be produced in shorter or longer than 10-foot lengths, with or without belled ends. Where subject to physical damage for above-ground or underground applications, RNMC Schedule 80 extra-heavy-wall PVC-80 *must be used*. PVC Schedule 80 is also produced in ½ (16) inch through 6 (155) inch sizes, with or without belled ends. *NEC*® Article 347 permits rigid nonmetallic conduit and fittings to be used under the following conditions (an informational FPN advises that extreme cold can cause some nonmetallic conduits to become brittle and therefore more susceptible to damage from physical contact).

RNMC is permitted to be installed concealed in walls, floors, and ceilings in locations subject to severe corrosive influences, as covered by *NEC*® Section 300-6 or where subject to chemicals for which the materials are specifically approved. For chemicals for which PVC is not acceptable, the designer or installer should select an appropriate wiring method from Chapter 3 of the *NEC*®. **Note:** See Figure A–13 in the Appendix. The installations must comply with the specific *NEC*® article and, in addition, be installed in accordance

For underground applications encased in concrete or direct burial. Also for use in exposed or concealed applications aboveground.
- UL Listed
- Sunlight resistant
- Rated for use with 90°C conductors
- Superior weathering characteristics

For use in aboveground and belowground applications that are subject to physical damage.
- UL Listed
- Sunlight resistant
- Rated for use with 90°C conductors
- Superior weathering characteristics

Figure 14–28 Installation as per *NEC*® Article 347 rigid nonmetallic conduit. (Courtesy of Carlon, a Lampson Sessions Company.)

with the specific listing instructions as applicable. Uses vary by UL listing. Some non-metallic conduits are listed and marked for underground use only.

In wet locations, such as dairies, laundries, canneries, and other wet locations or where walls are frequently washed, the entire conduit system shall be so installed and equipped as to prevent water from entering the system. All supports, bolts, straps, screws, and so on shall be of a corrosion-resistant material or protected against corrosion by corrosion-resistant materials. RNMC can be used in dry and damp locations and exposed where not subject to physical damage, except Schedule 80 can be installed in locations subject to physical damage. Rigid nonmetallic conduit can be installed underground as per *NEC*® Sections 300-5 and 300-50. **Note:** Schedule 80 has less interior space and a reduction in wire fill is necessary. See *NEC*® Chapter 9, Tables 3A through 3F. Underground installations are to be made, as per *NEC*® Sections 300-5 and 300-50.

Rigid nonmetallic conduit is not permitted to be used in hazardous locations except as per the *NEC*®. It is not permitted for the support of fixtures or other equipment unless identified specifically for such use. It is not permitted where subject to ambient temperatures exceeding 50°C (122°F). It is not permitted in theaters or similar locations, except under limited conditions as provided in *NEC*® Articles 518 and 520. Rigid nonmetallic conduit is not permitted to be used in environmental airspaces as detailed in *NEC*® Article 300-22.

Rigid nonmetallic conduit products are joined together by means of solvent cement. Sizes ½ inch through 1½ inches should be cut square using a fine-toothed saw and deburred. For sizes 2 inches through 6 inches, a miter box or similar saw guide should be used to keep the material steady. After cutting and deburring, wipe the ends clean of dust, dirt, and shavings. The joining process is as follows: Be sure that the conduit end is clean and dry. Place a coating of primer/cleaner on the end and then on its mating part with a dauber. Thoroughly coat the surfaces to be mated. Allow primer/cleaner a few seconds to soften the PVC surface (the time might need to be adjusted, depending on the temperature). Apply a coat of solvent cement to the end of the conduit for the length of the socket to be attached. Push conduit firmly into fitting while rotating conduit slightly about one-quarter turn to spread cement evenly. Allow the joint to set approximately 10 minutes. Most manufacturers recommend specific solvent cement. The cement is prepared for their products, compounds, and tolerances, and substitutions should not be made as adverse effects can occur. In situations requiring extremely fast setting or in low temperature or difficult installation conditions, all-weather quick-set cement is available and should be used. Rigid nonmetallic conduit can be bent in the field. *NEC*® Sections 347-13 and 347-15 specify how the bends are to be made and the number permitted in a run. They must be made so that the conduit is not damaged and the internal diameter of the conduit will not effectively be reduced. Bends shall be made only with bending equipment identified for the purpose. For more information on bending, see Chapter 3. The number of conductors permitted shall not exceed that permitted by the percentage fill in *NEC*® Table 1, Chapter 9.

GLOSSARY

ANSI: American National Standards Institute; ANSI is an umbrella organization for other standards-writing organizations. After a standard has been developed and gone through the consensus process, it is then processed as an American National Standard.

API: American Petroleum Institute; they have unique recommendations for the classification of areas dealing with petroleum facilities, such as API IRP 500.

ASTM E119: Standard Method of Fire Test for Building Construction and Materials.

Authority having jurisdiction (AHJ): The person or discipline responsible for enforcement of the *NEC®* and having the responsibility for making interpretations of the rules, for deciding on the approval of equipment and materials, and for granting the special permission contemplated in a number of the rules.

BOCA: Building Officials and Code Administrators International, Country Club Hills, Illinois; the national codes are adopted throughout the eastern and upper Midwest. They produce the National Building Code, Plumbing, Mechanical, Fire, and so on.

Consensus process: All changes are made in a democratic fashion on the basis of public input and voted and acted on by a balanced committee of voluntary experts or representatives from all facets of the standard.

Fine-print notes (FPNs): Explanatory information and references are in the form of fine-print notes, which are identified (FPN). These FPNs are not enforceable.

Flash point: The minimum temperature at which a liquid gives off vapor in sufficient concentration to form an ignitable mixture of air near the surface of the liquid within the vessel as specified by an appropriate test procedure and apparatus.

Example: The flash point of a liquid having a viscosity less than 45 SUS at 100°F (37.8°C) and a flash point below 200°F (93.4°C) is determined in accordance with ASTM D56, Standard Method of Test for Flash Point by the TAG Closed Tester. (See definition of flash point in NFPA 30A or other NFPA standards for flammable materials.)

GRC: Rigid metal conduit as covered in *NEC®* Article 346 galvanized steel conduit.

Green Book: The UL *Electrical Constructions Material Directory.*

Hazardous atmosphere: Atmospheres containing concentrations of flammable vapors, gases, or liquids or combustible dusts or ignitable fibers or flyings.

ICBO: International Conference of Building Officials, Whittier, California. The uniform codes are adopted throughout the western and midwestern states. They produce all of the uniform codes, such as the building, mechanical, plumbing, and fire codes.

ID: Internal diameter.

IEC: International Electrotechnical Commission, based in Europe; writes standards for international use. Often thought of as a European organization; however, the United States actively participates in this standards-writing body along with representatives from many other countries.

Label: The information label on the product, usually containing the listing mark.

Listing mark: Often called the "bug" or trademark, generally the listing agency's initials in a unique configuration.

National Electrical Code® (NEC®): The most widely used electrical code in the world is the *NEC®*. The official designation for the *National Electrical Code®* is ANSI/NFPA 70.

NEMA: The National Electrical Manufacturers Association located in Rosslyn, Virginia.

NFPA: National Fire Protection Association; a standards-writing organization located in Quincy, Massachusetts, that is responsible for many fire and safety standards.

Nonincendive circuits: A circuit in which any arc or thermal effect produced, under intended operating conditions of the equipment or due to the opening, shorting, or grounding of the field wiring, is not capable, under specified test conditions, of igniting the flammable gas, vapor, or dust-air mixture.

NRTL: Nationally Recognized Testing Laboratory. These are laboratories that are recognized by OSHA as competent testing laboratories.

OD: Outside diameter.

OSHA: Occupational Safety and Health Administration (Act), intended to protect employees from hazards in the workplace.

Red Book: *Hazardous Location Equipment* will provide valuable information that can help avoid the misapplication of materials.

RMC: Rigid metal conduit as covered in *NEC®* Article 346; usually galvanized steel or aluminum but can be other metals, such as stainless steel or brass.

SBCCI: Southern Building Code Congress International Inc., Birmingham, Alabama. The standard codes are adopted throughout the southern states. They produce all the standard codes and building, mechanical, plumbing, and fire codes.

White Book: General Information for *Electrical Construction, Hazardous Location, and Electric Heating and Air-Conditioning Equipment.*

APPENDIX

Hazardous Substances Tables

Class 1* Group	Substance	Auto-* Ignition Temp. °F	Auto-* Ignition Temp. °C	Flash** Point °F	Flash** Point °C	Explosive Limits** Per Cent by Volume Lower	Explosive Limits** Per Cent by Volume Upper	Vapor** Density (Air Equals 1.0)
C	Acetaldehyde	347	175	-38	-39	4.0	60	1.5
D	Acetic Acid	867	464	103	39	4.0	19.9 @ 200° F	2.1
D	Acetic Anhydride	600	316	120	49	2.7	10.3	3.5
D	Acetone	869	465	-4	-20	2.5	13	2.0
D	Acetone Cyanohydrin	1270	688	165	74	2.2	12.0	2.9
D	Acetonitrile	975	524	42	6	3.0	16.0	1.4
A	Acetylene	581	305	gas	gas	2.5	100	0.9
B(C)	Acrolein (inhibited)[1]	455	235	-15	-26	2.8	31.0	1.9
D	Acrylic Acid	820	438	122	50	2.4	8.0	2.5
D	Acrylonitrile	898	481	32	0	3.0	17	1.8
D	Adiponitrile	—	—	200	93	—	—	2.0
C	Allyl Alcohol	713	378	70	21	2.5	18.0	2.0
D	Allyl Chloride	905	485	-25	-32	2.9	11.1	2.6
B(C)	Allyl Glycidyl Ether[1]	—	—	—	—	—	—	0.6
D	Ammonia[2]	928	498	gas	gas	15	28	
D	n-Amyl Acetate	680	360	60	16	1.1	7.5	4.5
D	sec-Amyl Acetate			89	32	—	—	4.5
D	Aniline	1139	615	158	70	1.3	11	3.2
D	Benzene	928	498	12	-11	1.3	7.9	2.8
D	Benzyl Chloride	1085	585	153	67	1.1	—	4.4
B(D)	1,3-Butadiene[1]	788	420	gas	gas	2.0	12.0	1.9
D	Butane	550	288	gas	gas	1.6	8.4	2.0
D	1-Butanol	650	343	98	37	1.4	11.2	2.6
D	2-Butanol	761	405	75	24	1.7 @ 212° F	9.8 @ 212° F	2.6
D	n-Butyl Acetate	790	421	72	22	1.7	7.6	4.0
D	iso-Butyl Acetate	790	421	—	—	—	—	4.0
D	sec-Butyl Acetate	—	—	88	31	1.7	9.8	—
D	t-Butyl Acetate	—	—	—	—	—	—	—
D	n-Butyl Acrylate (inhibited)	559	293	118	48	1.5	9.9	4.4
C	n-Butyl Formal	—	—	—	—	—	—	—
B(C)	n-Butyl Glycidyl Ether[1]	—	—	35	2	—	—	3.1
C	Butyl Mercaptan	—	—	—	—	—	—	—
C	t-Butyl Toluene	—	—	10	-12	1.7	9.8	2.5
D	Butylamine	594	312					
D	Butylene	725	385	gas	gas	1.6	10.0	1.9
C	n-Butyraldehyde	425	218	-8	-22	1.9	12.5	2.5
D	n-Butyric Acid	830	443	161	72	2.0	10.0	3.0
[3]	Carbon Disulfide	194	90	-22	-30	1.3	50.0	2.6
C	Carbon Monoxide	1128	609	gas	gas	12.5	74.0	1.0
C	Chloroacetaldehyde	—	—	—	—	—	—	—
D	Chlorobenzene	1099	593	82	28	1.3	9.6	3.9
C	1-Chloro-1-Nitropropane	—	—	144	62	—	—	4.3
D	Chloroprene	—	—	-4	-20	4.0	20.0	3.0
D	Cresol	1038-1110	559-599	178-187	81-86	1.1-1.4	—	—
C	Crotonaldehyde	450	232	55	13	2.1	15.5	2.4
D	Cumene	795	424	96	36	0.9	6.5	4.1
D	Cyclohexane	473	245	-4	-20	1.3	8.0	2.9
D	Cyclohexanol	572	300	154	68	—	—	3.5
D	Cyclohexanone	473	245	111	44	1.1 @ 212° F	9.4	3.4
D	Cyclohexene	471	244	<20	<-7	—	—	2.8
D	Cyclopropane	938	503	gas	gas	2.4	10.4	1.5
D	p-Cymene	817	436	117	47	0.7 @ 212° F	5.6	4.6
C	n-Decaldehyde	—	—	—	—	—	—	—
D	n-Decanol	550	288	180	82	—	—	5.5
D	Decene	455	235	<131	<55	—	—	4.84
D	Diacetone Alcohol	1118	603	148	64	1.8	6.9	4.0
D	o-Dichlorobenzene	1198	647	151	66	2.2	9.2	5.1
D	1,1-Dichloroethane	820	438	22	-6	5.6	—	—
D	1,2-Dichloroethylene	860	460	36	2	5.6	12.8	3.4
C	1,1-Dichloro-1-Nitroethane	—	—	168	76	—	—	5.0
D	1,3-Dichloropropene	—	—	95	35	5.3	14.5	3.8
C	Dicyclopentadiene	937	503	90	32	—	—	—
D	Diethyl Benzene	743-842	395-450	133-135	56-57	—	—	4.6
C	Diethyl Ether	320	160	-49	-45	1.9	36.0	2.6
C	Diethylamine	594	312	-9	-23	1.8	10.1	2.5
C	Diethylaminoethanol	—	—	—	—	—	—	—
C	Diethylene Glycol Monobutyl Ether	442	228	172	78	0.85	24.6	5.6
C	Diethylene Glycol Monomethyl Ether	465	241	205	96	—	—	—
D	Di-isobutyl Ketone	745	396	120	49	0.8 @ 200° F	7.1 @ 200° F	4.9
D	Di-isobutylene	736	391	23	-5	0.8	4.8	3.9
C	Di-isopropylamine	600	316	30	-1	1.1	7.1	3.5
D	N-N-Dimethyl Aniline	700	371	145	63	—	—	4.2
D	Dimethyl Formamide	833	455	136	58	2.2 @ 212° F	15.2	2.5
D	Dimethyl Sulfate	370	188	182	83	—	—	4.4
C	Dimethylamine	752	400	gas	gas	2.8	14.4	1.6
C	1,4-Dioxane	356	180	54	12	2.0	22	3.0
D	Dipentene	458	237	113	45	0.7 @ 302° F	6.1 @ 302° F	4.7
C	Di-n-propylamine	570	299	63	17	—	—	3.5
C	Dipropylene Glycol Methyl Ether	—	—	185	85	—	—	5.11
D	Dodecene	491	255	—	—	—	—	—

Figure A–1(A) Properties of gases, liquids, and vapors. (Courtesy of Crouse-Hinds, Division of Cooper Industries.)

Class 1* Group	Substance	Auto-Ignition Temp. °F	Auto-Ignition Temp. °C	Flash Point °F	Flash Point °C	Explosive Limits Lower	Explosive Limits Upper	Vapor Density (Air Equals 1.0)
C	Epichlorohydrin	772	411	88	31	3.8	21.0	3.2
D	Ethane	882	472	gas	gas	3.0	12.5	1.0
D	Ethanol	685	363	55	13	3.3	19	1.6
D	Ethyl Acetate	800	427	24	-4	2.0	11.5	3.0
D	Ethyl Acrylate (inhibited)	702	372	50	10	1.4	14	3.5
D	Ethyl sec-Amyl Ketone	—	—	—	—	—	—	—
D	Ethyl Benzene	810	432	70	21	1.0	6.7	3.7
D	Ethyl Butanol	—	—	—	—	—	—	—
D	Ethyl Butyl Ketone	—	—	115	46	—	—	4.0
D	Ethyl Chloride	966	519	-58	-50	3.8	15.4	2.2
D	Ethyl Formate	851	455	-4	-20	2.8	16.0	2.6
D	2-Ethyl Hexanol	448	231	164	73	0.88	9.7	4.5
D	2-Ethyl Hexyl Acrylate	485	252	180	82	—	—	—
C	Ethyl Mercaptan	572	300	<0	<-18	2.8	18.0	2.1
C	n-Ethyl Morpholine	—	—	—	—	—	—	—
C	2-Ethyl-3-Propyl Acrolein	—	—	155	68	—	—	4.4
D	Ethyl Silicate	—	—	125	52	—	—	7.2
D	Ethylamine	725	385	<0	<-18	3.5	14.0	1.6
C	Ethylene	842	450	gas	gas	2.7	36.0	1.0
D	Ethylene Chlorohydrin	797	425	140	60	4.9	15.9	2.8
D	Ethylene Dichloride	775	413	56	13	6.2	16	3.4
C	Ethylene Glycol Monobutyl Ether	460	238	143	62	1.1 @ 200° F	12.7 @ 275° F	4.1
C	Ethylene Glycol Monobutyl Ether Acetate	645	340	160	71	0.88 @ 200° F	8.54 @ 275° F	—
C	Ethylene Glycol Monoethyl Ether	455	235	110	43	1.7 @ 200° F	15.6 @ 200° F	3.0
C	Ethylene Glycol Monoethyl Ether Acetate	715	379	124	52	1.7	—	4.72
D	Ethylene Glycol Monomethyl ether	545	285	102	39	1.8 @ STP	14 @ STP	2.6
B(C)	Ethylene Oxide'	804	429	-20	-28	3.0	100	1.5
D	Ethylenediamine	725	385	93	34	4.2	14.4	2.1
C	Ethylenimine	608	320	12	-11	3.6	46.0	1.5
C	2-Ethylhexaldehyde	375	191	112	44	0.85 @ 200° F	7.2 @ 275° F	4.4
B	Formaldehyde (Gas)	795	429	gas	gas	7.0	73	1.0
D	Formic Acid (90%)	813	434	122	50	18	57	—
D	Fuel Oils	410-765	210-407	100-336	38-169	0.7	5	—
C	Furfural	600	316	140	60	2.1	19.3	3.3
C	Furfuryl Alcohol	915	490	167	75	1.8	16.3	3.4
D	Gasoline	536-880	280-471	-36 to -50	-38 to -46	1.2-1.5	7.1-7.6	3-4
D	Heptane	399	204	25	-4	1.05	6.7	3.5
D	Heptene	500	260	<32	<0	—	—	3.39
D	Hexane	437	225	-7	-22	1.1	7.5	3.0
D	Hexanol	—	—	145	63	—	—	3.5
D	2-Hexanone	795	424	77	25	—	8	3.5
D	Hexenes	473	245	<20	<-7	—	—	3.0
D	sec-Hexyl Acetate	—	—	—	—	—	—	—
C	Hydrazine	74-518	23-270	100	38	2.9	9.8	1.1
B	Hydrogen	968	520	gas	gas	4.0	75	0.1
C	Hydrogen Cyanide	1000	538	0	-18	5.6	40.0	0.9
C	Hydrogen Selenide	—	—	—	—	—	—	—
C	Hydrogen Sulfide	500	260	gas	gas	4.0	44.0	1.2
D	Isoamyl Acetate	680	360	77	25	1.0 @ 212° F	7.5	4.5
D	Isoamyl Alcohol	662	350	109	43	1.2	9.0 @ 212° F	3.0
D	Isobutyl Acrylate	800	427	86	30	—	—	4.42
C	Isobutyraldehyde	385	196	-1	-18	1.6	10.6	2.5
C	Isodecaldehyde	—	—	185	85	—	—	5.4
D	Iso-octyl Alcohol	—	—	180	82	—	—	—
C	Iso-octyl Aldehyde	387	197	—	—	—	—	—
D	Isophorone	860	460	184	84	0.8	3.8	—
D	Isoprene	428	220	-65	-54	1.5	8.9	2.4
D	Isopropyl Acetate	860	460	35	2	1.8 @ 100° F	8	3.5
D	Isopropyl Ether	830	443	-18	-28	1.4	7.9	3.5
C	Isopropyl Glycidyl Ether	—	—	—	—	—	—	—
D	Isopropylamine	756	402	-35	-37	—	—	2.0
D	Kerosene	410	210	110-162	43-72	0.7	5	—
D	Liquefied Petroleum Gas	761-842	405-450	—	—	—	—	—
B	Manufactured Gas (containing more than 30% H₂ by volume)	—	—	—	—	—	—	—
D	Mesityl Oxide	652	344	87	31	1.4	7.2	3.4
D	Methane	999	630	gas	gas	5.0	15.0	0.6
D	Methanol	725	385	52	11	6.0	36	1.1
D	Methyl Acetate	850	454	14	-10	3.1	16	2.8
D	Methyl Acrylate	875	468	27	-3	2.8	25	3.0
D	Methyl Amyl Alcohol	—	—	106	41	1.0	5.5	—
D	Methyl n-Amyl Ketone	740	393	102	39	1.1 @ 151° F	7.9 @ 250° F	3.9
C	Methyl Ether	662	350	gas	gas	3.4	27.0	1.6
D	Methyl Ethyl Ketone	759	404	16	-9	1.7 @ 200° F	11.4 @ 200° F	2.5
D	2-Methyl-5-Ethyl Pyridine	—	—	155	68	1.1	6.6	4.2
C	Methyl Formal	460	238	—	—	—	—	—
D	Methyl Formate	840	449	-2	-19	4.5	23	2.1

Figure A–1(B) Properties of gases, liquids, and vapors. (Courtesy of Crouse-Hinds, Division of Cooper Industries.)

Class 1* Group	Substance	Auto-* Ignition Temp. °F	°C	Flash ** Point °F	°C	Explosive Limits** Per Cent by Volume Lower	Upper	Vapor** Density (Air Equals 1.0)
D	Methyl Isobutyl Ketone	840	440	64	18	1.2 @ 200° F	8.0 @ 200° F	**3.5**
D	Methyl Isocyanate	994	534	19	-7	5.3	26	**1.97**
C	Methyl Mercaptan	—	—	—	—	3.9	21.8	1.7
D	Methyl Methacrylate	792	422	50	10	1.7	8.2	**3.6**
D	2-Methyl-1-Propanol	780	416	82	28	1.7 @ 123° F	10.6 @ 202° F	**2.6**
D	2-Methyl-2-Propanol	892	478	52	11	2.4	8.0	**2.6**
D	alpha-Methyl Styrene	1066	574	129	54	1.9	6.1	—
C	Methylacetylene	—	—	gas	gas	1.7		1.4
C	Methylacetylene-Propadiene (stabilized)	—	—	—	—	—	—	—
D	Methylamine	806	430	gas	gas	4.9	20.7	1.0
D	Methylcyclohexane	482	250	25	-4	1.2	6.7	3.4
D	Methylcyclohexanol	565	296	149	65	—	—	3.9
D	o-Methylcyclohexanone	—	—	118	48	—	—	3.9
D	Monoethanolamine	770	410	185	85	—	—	2.1
D	Monoisopropanolamine	705	374	171	77	—	—	2.6
C	Monomethyl Aniline	900	482	185	85	—	—	3.7
C	Monomethyl Hydrazine	382	194	17	-8	2.5	92	1.6
C	Morpholine	590	310	98	37	1.4	11.2	3.0
D	Naphtha (Coal Tar)	531	277	107	42	—	—	—
D	Naphtha (Petroleum)⁴	550	288	<0	<-18	1.1	5.9	2.5
D	Nitrobenzene	900	482	190	88	1.8 @ 200° F	—	4.3
C	Nitroethane	778	414	82	28	3.4	—	2.6
C	Nitromethane	785	418	95	35	7.3	—	2.1
C	1-Nitropropane	789	421	96	36	2.2	—	3.1
C	2-Nitropropane	802	428	75	24	2.6	11.0	3.1
D	Nonane	401	205	88	31	0.8	2.9	4.4
D	Nonene	—	—	78	26	—	—	4.35
D	Nonyl Alcohol	—	—	165	74	0.8 @ 212° F	6.1 @ 212° F	5.0
D	Octane	403	206	56	13	1.0	6.5	3.9
D	Octene	446	230	70	21	—	—	3.9
D	n-Octyl Alcohol	—	—	178	81	—	—	4.5
D	Pentane	470	243	<-40	<-40	1.5	7.8	2.5
D	1-Pentanol	572	300	91	33	1.2	10.0 @ 212° F	3.0
D	2-Pentanone	846	452	45	7	1.5	8.2	3.0
D	1-Pentene	527	275	0	-18	1.5	8.7	2.4
D	Phenylhydrazine	—	—	190	88	—	—	—
D	Propane	842	450	gas	gas	2.1	9.5	1.6
D	1-Propanol	775	413	74	23	2.2	13.7	2.1
D	2-Propanol	750	399	53	12	2.0	12.7 @ 200° F	2.1
D	Propiolactone	—	—	165	74	2.9	—	2.5
C	Propionaldehyde	405	207	-22	-30	2.6	17	2.0
D	Propionic Acid	870	466	126	52	2.9	12.1	2.5
D	Propionic Anhydride	545	285	145	63	1.3	9.5	4.5
D	n-Propyl Acetate	842	450	55	13	1.7 @ 100° F	8	3.5
C	n-Propyl Ether	419	215	70	21	1.3	7.0	3.53
B	Propyl Nitrate	347	175	68	20	2	100	—
D	Propylene	851	455	gas	gas	2.0	11.1	1.5
D	Propylene Dichloride	1035	557	60	16	3.4	14.5	3.9
B(C)	Propylene Oxide¹	840	449	-35	-37	2.3	36	2.0
D	Pyridine	900	482	68	20	1.8	12.4	2.7
D	Styrene	914	490	88	31	1.1	7.0	3.6
C	Tetrahydrofuran	610	321	6	-14	2.0	11.8	2.5
D	Tetrahydronaphthalene	725	385	160	71	0.8 @ 212° F	5.0 @ 302° F	4.6
C	Tetramethyl Lead	—	—	100	38	—	—	6.5
D	Toluene	896	480	40	4	1.2	7.1	3.1
D	Tridecene	—	—	—	—	—	—	—
C	Triethylamine	480**	249**	16	-9	1.2	8.0	3.5
D	Triethylbenzene	—	—	181	83	—	—	5.6
D	Tripropylamine	—	—	105	41	—	—	4.9
D	Turpentine	488	253	95	35	0.8	—	—
D	Undecene	—	—	—	—	—	—	—
C	Unsymmetrical Dimethyl Hydrazine (UDMH)	480	249	5	-15	2	95	2.0
C	Valeraldehyde	432	222	54	12	—	—	3.0
D	Vinyl Acetate	756	402	18	-8	2.6	13.4	3.0
D	Vinyl Chloride	882	472	gas	gas	3.6	33.0	2.2
D	Vinyl Toluene	921	494	120	49	—	11.0	4.1
D	Vinylidene Chloride	1058	570	-19	-28	6.5	15.5	3.4
D	Xylenes	867-984	464-529	81-90	27-32	1.0-1.1	7.0	3.7

¹ If equipment is isolated by sealing all conduit ½ in. or larger, in accordance with Section 501-5(a) of NFPA 70, *National Electrical Code*, equipment for the group classification shown in parentheses is permitted.

² For classification of areas involving Ammonia, see *Safety Code for Mechanical Refrigeration*, ANSI/ASHRAE 15, and *Safety Requirements for the Storage and Handling of Anhydrous Ammonia*, ANSI/CGA G2.1.

³ Certain chemicals may have characteristics that require safeguards beyond those required for any of the above groups. Carbon disulfide is one of these chemicals because of its low autoignition temperature and the small joint clearance to arrest its flame propagation.

⁴ Petroleum Naphtha is a saturated hydrocarbon mixture whose boiling range is 20° to 135° C. It is also known as benzine, ligroin, petroleum ether, and naphtha.

˙ Data from NFPA 497M-1986, *Classification of Gases, Vapors and Dusts for Electrical Equipment in Hazardous (Classified) Locations*.

˙˙ Data from NFPA 325M-1984, *Fire Hazard Properties of Flammable Liquids, Gases and Volatile Solids*.

Figure A–1(C)   Properties of gases, liquids, and vapors. (Courtesy of Crouse-Hinds, Division of Cooper Industries.)

Class II, Group E Material[2]	°F		°C
Aluminum, atomized collector fines	1022	Cl	550
Aluminum, A422 flake	608		320
Aluminum — cobalt alloy (60-40)	1058		570
Aluminum — copper alloy (50-50)	1526		830
Aluminum — lithium alloy (15% Li)	752		400
Aluminum — magnesium alloy (Dowmetal)	806	Cl	430
Aluminum — nickel alloy (58-42)	1004		540
Aluminum — silicon alloy (12% Si)	1238	NL	670
Boron, commercial-amorphous (85% B)	752		400
Calcium Silicide	1004		540
Chromium, (97%) electrolytic, milled	752		400
Ferromanganese, medium carbon	554		290
Ferrosilicon (88%, 9% Fe)	1472		800
Ferrotitanium (19% Ti, 74.1% Fe, 0.06% C)	698	Cl	370
Iron, 98%, H_2 reduced	554		290
Iron, 99%, Carbonyl	590		310
Magnesium, Grade B, milled	806		430
Manganese	464		240
Silicon, 96%, milled	1436	Cl	780
Tantalum	572		300
Thorium, 1.2%, O_2	518	Cl	270
Tin, 96%, atomized (2% Pb)	806		430
Titanium, 99%	626	Cl	330
Titanium Hydride, (95% Ti, 3.8% H_2)	896	Cl	480
Vanadium, 86.4%	914		490
Zirconium Hydride, (93.6% Zr, 2.1% H_2)	518		270

Class II, Group G
AGRICULTURAL DUSTS

Material	°F		°C
Alfalfa Meal	392		200
Almond Shell	392		200
Apricot Pit	446		230
Cellulose	500		260
Cherry Pit	428		220
Cinnamon	446		230
Citrus Peel	518		270
Cocoa Bean Shell	698		370
Cocoa, natural, 19% fat	464		240
Coconut Shell	428		220
Corn	482		250
Corncob Grit	464		240
Corn Dextrine	698		370
Cornstarch, commercial	626		330
Cornstarch, modified	392		200
Cork	410		210
Cottonseed Meal	392		200
Cube Root, South Amer.	446		230
Flax Shive	446		230
Garlic, dehydrated	680	NL	360
Guar Seed	932	NL	500
Gum, Arabic	500		260
Gum, Karaya	464		240
Gum, Manila (copal)	680	Cl	360
Gum, Tragacanth	500		260
Hemp Hurd	428		220
Lycopodium	590		310
Malt Barley	482		250
Milk, Skimmed	392		200
Pea Flour	500		260
Peach Pit Shell	410		210
Peanut Hull	410		210
Peat, Sphagnum	464		240
Pecan Nut Shell	410		210
Pectin	392		200
Potato Starch, Dextrinated	824	NL	440
Pyrethrum	410		210
Rauwolfia Vomitoria Root	446		230
Rice	428		220
Rice Bran	914	NL	490
Rice Hull	428		220
Safflower Meal	410		210
Soy Flour	374		190
Soy Protein	500		260
Sucrose	662	Cl	350
Sugar, Powdered	698	Cl	370

Class II, Group G (cont'd)
AGRICULTURAL DUSTS

Material	°F		°C
Tung, Kernels, Oil-Free	464		240
Walnut Shell, Black	428		220
Wheat	428		220
Wheat Flour	680		360
Wheat Gluten, gum	968	NL	520
Wheat Starch	716	NL	380
Wheat Straw	428		220
Woodbark, Ground	482		250
Wood Flour	500		260
Yeast, Torula	500		260

CARBONACEOUS DUSTS[3]

Material	°F		°C
Asphalt, (Blown Petroleum Resin)	950	Cl	510
Charcoal	356		180
Coal, Kentucky Bituminous	356		180
Coal, Pittsburgh Experimental	338		170
Coal, Wyoming	—		—
Gilsonite	932		500
Lignite, California	356		180
Pitch, Coal Tar	1310	NL	710
Pitch, Petroleum	1166	NL	630
Shale, Oil	—		—

CHEMICALS

Material	°F		°C
Acetoacetanilide	824	M	440
Acetoacet-p-phenetidide	1040	NL	560
Adipic Acid	1022	M	550
Anthranilic Acid	1076	M	580
Aryl-nitrosomethylamide	914	NL	490
Azelaic Acid	1130	M	610
2,2-Azo-bis-butyronitrile	662		350
Benzoic Acid	824	M	440
Benzotriazole	824	M	440
Bisphenol-A	1058	M	570
Chloroacetoacetanilide	1184	M	640
Diallyl Phthalate	896	M	480
Dicumyl Peroxide (suspended on $CaCO_3$), 40-60	356		180
Dicyclopentadiene Dioxide	788	NL	420
Dihydroacetic Acid	806	NL	430
Dimethyl Isophthalate	1076	M	580
Dimethyl Terephthalate	1058	M	570
3,5 - Dinitrobenzoic Acid	860	M	460
Dinitrotoluamide	932	NL	500
Diphenyl	1166	M	630
Ditertiary Butyl Paracresol	878	NL	470
Ethyl Hydroxyethyl Cellulose	734	NL	390
Fumaric Acid	968	M	520
Hexamethylene Tetramine	770	S	410
Hydroxyethyl Cellulose	770	NL	410
Isotoic Anhydride	1292	NL	700
Methionine	680		360
Nitrosoamine	518	NL	270
Para-oxy-benzaldehyde	716	Cl	380
Paraphenylene Diamine	1148	M	620
Paratertiary Butyl Benzoic Acid	1040	M	560
Pentaerythritol	752	M	400
Phenylbetanaphthylamine	1256	NL	680
Phthalic Anydride	1202	M	650
Phthalimide	1166	M	630
Salicylanilide	1130	M	610
Sorbic Acid	860		460
Stearic Acid, Aluminum Salt	572		300
Stearic Acid, Zinc Salt	950	M	510
Sulfur	428		220
Terephthalic Acid	1256	NL	680

DRUGS

Material	°F		°C
2-Acetylamino-5-nitrothiazole	842		450
2-Amino-5-nitrothiazole	860		460
Aspirin	1220	M	660
Gulasonic Acid, Diacetone	788	NL	420
Mannitol	860	M	460
Nitropyridone	806	M	430
1-Sorbose	698	M	370
Vitamin B1, mononitrate	680	NL	360
Vitamin C (Ascorbic Acid)	536		280

Figure A–2(A) Properties of combustible dusts. (Courtesy of Crouse-Hinds, Division of Cooper Industries.)

Class II, Group G	Minimum Cloud or Layer Ignition Temp.[1] °F		°C
DYES, PIGMENTS, INTERMEDIATES			
Beta-naphthalene-azo-Dimethylaniline	347		175
Green Base Harmon Dye	347		175
Red Dye Intermediate	347		175
Violet 200 Dye	347		175
PESTICIDES			
Benzethonium Chloride	716	CI	380
Bis(2-Hydroxy-5-chlorophenyl) methane	1058	NL	570
Crag No. 974	590	CI	310
Dieldrin (20%)	1022	NL	550
2, 6-Ditertiary-butyl-paracresol	788	NL	420
Dithane	356		180
Ferbam	302		150
Manganese Vancide	248		120
Sevin	284		140
α,α - Trithiobis (N,N-Dimethylthio-formamide)	446		230
THERMOPLASTIC RESINS AND MOLDING COMPOUNDS			
Acetal Resins			
Acetal, Linear (Polyformaldehyde)	824	NL	440
Acrylic Resins			
Acrylamide Polymer	464		240
Acrylonitrile Polymer	860		460
Acrylonitrile - Vinyl Pyridine Copolymer	464		240
Acrylonitrile-Vinyl Chloride-Vinylidene Chloride Copolymer (70-20-10)	410		210
Methyl Methacrylate Polymer	824	NL	440
Methyl Methacrylate - Ethyl Acrylate Copolymer	896	NL	480
Methyl Methacrylate-Ethyl Acrylate-Styrene Copolymer	824	NL	440
Methyl Methacrylate-Styrene-Butadiene-Acrylonitrile Copolymer	896	NL	480
Methacrylic Acid Polymer	554		290
Cellulosic Resins			
Cellulose Acetate	644		340
Cellulose Triacetate	806	NL	430
Cellulose Acetate Butyrate	698	NL	370
Cellulose Propionate	860	NL	460
Ethyl Cellulose	608	CI	320
Methyl Cellulose	644		340
Carboxymethyl Cellulose	554		290
Hydroxyethyl Cellulose	644		340
Chlorinated Polyether Resins			
Chlorinated Polyether Alcohol	860		460
Nylon (Polyamide) Resins			
Nylon Polymer (Polyhexa-methylene Adipamide)	806		430
Polycarbonate Resins			
Polycarbonate	1310	NL	710
Polyethylene Resins			
Polyethylene, High Pressure Process	716		380
Polyethylene, Low Pressure Process	788	NL	420
Polyethylene Wax	752	NL	400
Polymethylene Resins			
Carboxypolymethylene	968	NL	520

Class II, Group G (cont'd)	Minimum Cloud or Layer Ignition Temp. °F		°C
THERMOPLASTIC RESINS AND MOLDING COMPOUNDS			
Polypropylene Resins			
Polypropylene (No Antioxidant)	788	NL	420
Rayon Resins			
Rayon (Viscose) Flock	482		250
Styrene Resins			
Polystyrene Molding Cmpd.	1040	NL	560
Polystyrene Latex	932		500
Styrene-Acrylonitrile (70-30)	932	NL	500
Styrene-Butadiene Latex (> 75% Styrene; Alum Coagulated)	824	NL	440
Vinyl Resins			
Polyvinyl Acetate	1022	NL	550
Polyvinyl Acetate/Alcohol	824		440
Polyvinyl Butyral	734	NL	390
Vinyl Chloride - Acrylonitrile Copolymer	878		470
Polyvinyl Chloride - Dioctyl Phthalate Mixture	608	NL	320
Vinyl Toluene - Acrylonitrile Butadiene Copolymer	936	NL	530
THERMOSETTING RESINS AND MOLDING COMPOUNDS			
Allyl Resins			
Allyl Alcohol Derivative (CR-39)	932	NL	500
Amino Resins			
Urea Formaldehyde Molding Compound	860	NL	460
Urea Formaldehyde - Phenol Formaldehyde Molding Compound (Wood Flour Filler)	464		240
Epoxy Resins			
Epoxy	1004	NL	540
Epoxy - Bisphenol A	950	NL	510
Phenol Furfural	590		310
Phenolic Resins			
Phenol Formaldehyde	1076	NL	580
Phenol Formaldehyde Molding Cmpd (Wood Flour Filler)	932	NL	500
Phenol Formaldehyde, Polyalkylene-Polyamine Modified	554		290
Polyester Resins			
Polyethylene Terephthalate	932	NL	500
Styrene Modified Polyester-Glass Fiber Mixture	680		360
Polyurethane Resins			
Polyurethane Foam, No Fire Retardant	824		440
Polyurethane Foam, Fire Retardant	734		390
SPECIAL RESINS AND MOLDING COMPOUNDS			
Alkyl Ketone Dimer Sizing Compound	320		160
Cashew Oil, Phenolic, Hard	356		180
Chlorinated Phenol	1058	NL	570
Coumarone-Indene, Hard	968	NL	520
Ethylene Oxide Polymer	662	NL	350
Ethylene-Maleic Anhydride Copolymer	1004	NL	540
Lignin, Hydrolized, Wood-Type, Fines	842	NL	450
Petrin Acrylate Monomer	428	NL	220
Petroleum Resin (Blown Asphalt)	932		500
Rosin, DK	734	NL	390
Rubber, Crude, Hard	662	NL	350
Rubber, Synthetic, Hard (33% S)	608	NL	320
Shellac	752	NL	400
Sodium Resinate	428		220
Styrene — Maleic Anhydride Copolymer	878	CI	470

[1]Normally, the minimum ignition temperature of a layer of a specific dust is lower than the minimum ignition temperature of a cloud of that dust. Since this is not universally true, the lower of the two minimum ignition temperatures is listed. If no symbol appears between the two temperature columns, then the layer ignition temperature is shown. "CI" means the cloud ignition temperature is shown. "NL" means that no layer ignition temperature is available and the cloud ignition temperature is shown. "M" signifies that the dust layer melts before it ignites; the cloud ignition temperature is shown. "S" signifies that the dust layer sublimes before it ignites; the cloud ignition temperature is shown.

[2]Certain metal dusts may have characteristics that require safeguards beyond those required for atmospheres containing the dusts of aluminum, magnesium, and their commercial alloys. For example, zirconium, thorium, and uranium dusts have extremely low ignition temperatures (as low as 20° C) and minimum ignition energies lower than any material classified in any of the Class I or Class II groups.

[3]The 1987 NEC classifies carbonaceous dusts as Group F, some of which may be conductive.

Figure A–2(B) Properties of combustible dusts. (Courtesy of Crouse-Hinds, Division of Cooper Industries.)

USES FOR RACEWAY AND OTHER WIRING METHODS

CHAPTER 3, NEC®

Installation conditions

NEC ARTICLE	TYPE OF WIREWAY	For conductors over 600 volts	Inside Buildings	Outside Buildings	Under Ground	Cinder Fill	Embedded in Concrete	Wet locations	Dry Locations	Corrosive Locations	Severe Corrosive Locations	Hazardous Locations	Mechanical Injury	Severe Mechanical injury	Exposed work	Concealed work
330	TYPE MI MINERAL INSULATED-SHEATHED CABLE	X	P	P	C	X	P	P	P	C	X	P	P	P	P	P
331	ELECTRICAL NONMETALLIC TUBING*	X	C	X	X	X	C	P	P	P	C	X	X	X	C	C
333	TYPE AC ARMORED CABLE	X	P	X	X	X	X	P	X	X	X	X	X	X	C	P
334	TYPE MC METAL CLAD CABLE	C	C	C	C	X	C	C	P	X	X	C	X	X	P	P
336	NONMETALLIC SHEATHED CABLE ROMEX TYPE NM	X	P	C	X	X	X	X	P	X	X	X	X	X	P	P
338	TYPE SE SERVICE-ENTRANCE CABLE	X	C	P	X	X	X	C	C	X	X	X	X	X	C	C
388	TYPE USE SERVICE-ENTRANCE CABLE	X	X	C	P		X	C	X	X	X	X	X	X	X	X,
339	TYPE UF UNDERGROUND FEEDER AND BRANCH CIRCUIT CABLE	X	P	P	P		X	P	P	X	X	X	X	X	C	C
343	PREASSEMBLED CABLE IN NONMETALLIC CONDUIT	P	X	-	P	P		C	X	C	C	X	-	-	-	-
345	IMC INTERMEDIATE METAL CONDUIT	P	P	P	P	C	P	P	P	P	C	P	P	P	P	P
345	IMC INTERMEDIATE COATED METAL CONDUIT	P	C	P	P	P	P	P	P	P	P	P	P	P	P	P
346	RIGID METAL CONDUIT (STEEL)	P	P	P	P	C	P	P	P	P	C	P	P	P	P	P
346	RIGID COATED METAL CONDUIT	P	C	P	P	P	P	P	P	P	P	P	P	P	P	P
346	RIGID METAL CONDUIT (ALUMINUM)	P	P	P	X	X	X	P	P	X	X	P	P	C	P	P
347	RIGID NONMETALLIC CONDUIT O (SCHEDULE 40) *	C	P	P	P	P	P	P	P	P	P	X	X	X	C	P
347	RIGID NONMETALLIC CONDUIT (SCHEDULE 80) *	C	P	P	P	P	P	P	P	P	P	X	P	P	P	P
348	EMT ELECTRICAL METALLIC TUBING (STEEL)	C	P	P	P	C	P	P	P	C	P	C	P	X	P	P
349	FLEXIBLE METALLIC TUBING	C	C	X	X	X	X	X	C	X	X	X	X	X	C	C
350	FLEXIBLE METAL CONDUIT STEEL AND ALUMINUM	X	P	C	X	X	X	C	P	X	X	X	X	X	P	P
351A	LIQUIDTIGHT FLEXIBLE METAL CONDUIT	X	P	P	C	C	C	C	P	X	X	C	X	X	P	P
351b	LIQUIDTIGHT FLEXIBLE NONMETALLIC CONDUIT *	X	P	C	C	C	C	C	P	P	P	C	X	X	P	P
352A	SURFACE METAL RACEWAYS	C	P	C	-	-	-	C	P	C	C	C	C	C	P	-
352B	SURFACE NONMETALLIC RACEWAYS *	C	P	C	-	-	-	C	P	C	C	C	C	X	P	.
354	UNDERFLOOR RACEWAYS	C	P	-	-	-	C	X	P	X	X	C	C	C	-	P
356	CELLULAR METAL FLOOR RACEWAYS	C	P	-	-	-	P	X	P	-	-	—	-	-	.	-
362A	WIREWAYS METAL	C	P	C	X	X	X	P	P	X	X	C	P	X	P	X
362B	WIREWAYS NONMETALLIC	C	C	X	X	X	X	C	C	C	C	X	C	X	P	X
364	BUSWAYS	C	P	C	X	X	X	C	P	X	X	X	P	X	P	C

P - GENERALLY PERMITTED   C - CONDITIONAL   X - NOT PERMITTED   — CONDITIONS DO NOT APPLY

Wiring methods covered by *Chapter 3, NEC* ® are limited to systems utilizing voltages not exceeding 600 volts unless higher voltages are specifically permitted. *NEC* ® *300–21 & 22* may further limit some wiring methods.

* - Where subject to chemicals for which the raceway is specifically approved

* - (FPN) Extreme cold may cause some nonmetallic conduits to become brittle and therefore more susceptible to damage from physical contact

* - Except where the enclosed conductors' ambient temperature would exceed those for which the conduit is approved

Figure A–3   Chart showing recommended uses or applications for all types of wiring methods.

WEIGHTS AND DIMENSIONS FOR GALVANIZED RIGID CONDUIT

Trade Size, Inches	Approx. Wt. per 100 ft. (30.5m)		Nominal Outside Dia.[1]		Nominal Wall Thickness[2]		Length of Finished Conduit[3]		Quantity In Primary Bundle		Quantity In Master Bundle		Approx. Wt. of Master Bundle		Volume of Master Bundle	
	lb.	kg	in.	mm	in.	mm	ft.	m	ft.	m	ft.	m	lb.	kg	cu ft.	cu m
½	80	36.29	0.840	21.3	0.104	2.6	9'11¼"	3.03	100	30.48	2500	762	2000	907	20.8	0.59
¾	109	49.44	1.050	26.7	0.107	2.7	9'11¼"	3.03	50	15.24	2000	610	2180	989	24.3	0.69
1	165	74.84	1.315	33.4	0.126	3.2	9'11"	3.02	50	15.24	1250	381	2063	936	21.7	0.61
1¼	215	97.52	1.660	42.2	0.133	3.4	9'11"	3.02	30	9.14	900	274	1935	878	23.3	0.66
1½	258	117.03	1.900	48.3	0.138	3.5	9'11"	3.02	—	—	800	244	2064	936	27.8	0.79
2	352	159.67	2.375	60.3	0.146	3.7	9'11"	3.02	—	—	600	183	2112	958	33.8	0.96
2½	567	257.19	2.875	73.0	0.193	4.9	9'10½"	3.01	—	—	370	113	2098	952	29.2	0.83
3	714	323.87	3.500	88.9	0.205	5.2	9'10½"	3.01	—	—	300	91	2142	972	31.3	0.89
3½	860	390.10	4.000	101.6	0.215	5.5	9'10¼"	3.00	—	—	250	76	2150	975	34.7	0.98
4	1000	453.60	4.500	114.3	0.225	5.7	9'10¼"	3.00	—	—	200	61	2000	907	33.7	0.95
5	1320	598.75	5.563	141.3	0.245	6.2	9'10"	3.00	—	—	150	46	1980	898	41.3	1.17
6	1785	809.68	6.625	168.3	0.266	6.8	9'10"	3.00	—	—	100	30	1785	810	38.9	1.10

[1]Outside diameter tolerances: +/- .015 in. (.38mm) for trade sizes ½ in. through 2 in. +/- .025 in. (.64mm) for trade sizes 2½ in. through 4 in. +/- 1% for trade sizes 5 in. and 6 in.
[2]For more information only; not a spec requirement.
[3]Without Coupling Length Tolerances: +/- .25 in (6.35mm).

Figure A–4 Galvanized rigid conduit specifications. (Courtesy of Allied Tube & Conduit.)

WEIGHTS AND DIMENSIONS FOR INTERMEDIATE METAL CONDUIT

Trade Size, Inches	Approx. Wt. per 100 ft. (30.5m)		Nominal Outside Dia.[1]		Minimum Wall Thickness[2]		Length of Finished Conduit[3]		Quantity In Primary Bundle		Quantity In Master Bundle		Approx. Wt. of Master Bundle		Volume of Master Bundle	
	lb.	kg	in.	mm	in.	mm	ft.	m	ft.	m	ft.	m	lb.	kg	cu ft.	cu m
½	60	27.22	0.815	20.7	0.070	1.8	9'11¼"	3.03	100	30.48	3500	1067	2233	1013	26.7	0.76
¾	82	37.29	1.029	26.1	0.075	1.9	9'11¼"	3.03	50	15.24	2500	762	2078	943	30.7	0.87
1	116	52.62	1.290	32.8	0.085	2.2	9'11"	3.02	50	15.24	1700	518	2035	923	30.7	0.87
1¼	150	68.04	1.638	41.6	0.085	2.2	9'11"	3.02	—	—	1350	411	2209	1002	36.3	1.03
1½	182	82.55	1.883	47.8	0.090	2.3	9'11"	3.02	—	—	1100	335	2122	963	38.2	1.08
2	242	109.77	2.360	59.9	0.095	2.4	9'11"	3.02	—	--	800	244	2096	951	45.8	1.30
2½	428	194.14	2.857	72.6	0.140	3.5	9'10½"	3.01	—	—	370	113	1652	749	29.2	0.83
3	526	238.59	3.476	88.3	0.140	3.5	9'10½"	3.01	—	—	300	91	1618	734	31.3	0.89
3½	612	277.60	3.971	100.9	0.140	3.5	9'10¼"	3.00	—	—	240	73	1576	715	34.7	0.98
4	682	309.35	4.466	113.4	0.140	3.5	9'10¼"	3.00	—	—	240	73	1809	821	42.8	1.21

[1]Outside diameter tolerances: +/- .005 in. (.13mm) for trade sizes ½" through 1". +/- .0075 in. (.19mm) for trade size 1¼". through 2 in. +/- .010 in. (.25mm) for trade size 2½" through 4 in.
[2]Wall thickness tolerances: + .015 in. (.38mm) and - .000 for trade sizes ½" through 2 in. + 0.20 in. (.51 mm) and -.000 for trade sizes 2½" through 4 in.
Without Coupling Length Tolerances: +/- .25 in (6.35mm).

Figure A–5 Intermediate metal conduit specifications. (Courtesy of Allied Tube & Conduit.)

WEIGHTS AND DIMENSIONS FOR ELECTRICAL METALLIC TUBING (galvanized steel)

Trade Size, Inches	Approx. Wt. per 100 ft. (30.5m)		Nominal Outside Dia.¹		Nominal Wall Thickness		Length of Finished Conduit²		Quantity In Primary Bundle		Quantity In Master Bundle		Approx. Wt. of Master Bundle		Volume of Master Bundle	
	lb.	kg	in.	mm	in.	mm	ft.	m	ft.	m	ft.	m	lb.	kg	cu ft.	cu m
½	29	13.15	0.706	17.9	0.042	1.067	10	3.05	100	30.48	7000	2134	2037	924	28.7	0.81
¾	45	20.41	0.922	23.4	0.049	1.245	10	3.05	100	30.48	5000	1524	2175	987	35.6	1.01
1	65	29.48	1.163	29.5	0.057	1.448	10	3.05	50	15.24	3000	914	1905	864	33.7	0.95
1¼	96	43.55	1.510	38.4	0.065	1.651	10	3.05	50	15.24	2000	610	1894	859	35.0	0.99
1½	111	50.35	1.740	44.2	0.065	1.651	10	3.05	50	15.24	1500	457	1692	767	34.2	0.97
2	141	63.96	2.197	55.8	0.065	1.651	10	3.05	—	—	1200	366	1693	768	46.7	1.32
2½	215	97.52	2.875	73.0	0.072	1.829	10	3.05	—	—	610	186	1412	640	41.5	1.18
3	260	117.94	3.500	88.9	0.072	1.829	10	3.05	—	—	510	155	1429	648	48.9	1.38
3½	325	147.42	4.000	101.6	0.083	2.108	10	3.05	—	—	370	113	1248	566	48.6	1.38
4	390	176.90	4.500	114.3	0.083	2.108	10	3.05	—	—	300	91	1134	514	48.3	1.37

¹Outside diameter tolerances: +/- .005 in. (.13mm) for trade sizes ½" through 2". +/- .010 in. (.25mm) for trade size 2½". +/- .015 in. (.38mm) for trade size 3". +/- .020 in. (.51mm) for trade sizes 3½" and 4".
² Length tolerances: +/-.25" (6.35mm).

Figure A–6 Electrical metallic tubing (EMT) specifications. (Courtesy of Allied Tube & Conduit.)

PLUS 40® Heavy Wall

Nom. Size	Part No.	O.D.	I.D.	Wall	Wt. Per 100 Feet	Feet Per Bundle
½	49005	.840	.622	.109	18	100
¾	49007	1.050	.824	.113	23	100
1	49008	1.315	1.049	.133	35	100
1¼	49009	1.660	1.380	.140	48	50
1½	49010	1.900	1.610	.145	57	50
2	49011	2.375	2.067	.154	76	50
2½	49012	2.875	2.469	.203	125	10
3	49013	3.500	3.066	.216	164	10
3½	49014	4.000	3.548	.226	198	10
4	49015	4.500	4.026	.237	234	10
5	49016	5.563	5.047	.258	317	10
6	49017	6.625	6.065	.280	412	10

Rigid nonmetallic conduit is normally supplied in standard 10′ lengths, with one belled end per length. For specific requirements, it may be produced in lengths shorter or longer than 10′, with or without belled ends.

Figure A–7 Rigid nonmetallic conduit (Schedule 40) specifications. (Courtesy of Carlon, a Lampson Sessions Company.)

PLUS 80® Extra Heavy Wall

Nom. Size	Part No.	O.D.	I.D.	Wall	Wt. Per 100 Feet	Feet Per Bundle
$1/2$	49405	.840	.546	.147	21	100
$3/4$	49407	1.050	.742	.154	29	100
1	49408	1.315	.957	.179	43	100
$1 1/4$	49409	1.660	1.278	.191	60	50
$1 1/2$	49410	1.900	1.500	.200	72	50
2	49411	2.375	1.939	.218	100	10
$2 1/2$	49412	2.875	2.323	.276	153	10
3	49413	3.500	2.900	.300	212	10
4	49415	4.500	3.826	.337	310	10
5	49416	5.563	4.813	.375	431	10
6	49417	6.625	6.193	.432	592	10

Rigid nonmetallic conduit is normally supplied in standard 10′ lengths, with one belled end per length. For specific requirements, it may be produced in lengths shorter or longer than 10′, with or without belled ends.

Figure A–8   Rigid nonmetallic conduit (Schedule 80) specifications. (Courtesy of Carlon, a Lampson Sessions Company.)

Approximate Diameters for Conduit, Locknuts, and Bushings

Component Approximate diameter, inches	$1/2$	$3/4$	1	$1 1/4$	$1 1/2$	2	$2 1/2$	3	$3 1/2$	4	5	6
Conduit	$7/8$	$1 1/16$	$1 3/8$	$1 11/16$	$1 15/16$	$2 3/8$	$2 7/8$	$3 1/2$	4	$4 1/2$	$5 9/16$	$6 5/8$
Locknut	$1 1/8$	$1 3/8$	$1 11/16$	$2 3/16$	$2 7/16$	3	$3 7/16$	$4 3/16$	$5 3/8$	$6 11/16$	$7 15/16$	
Bushing	1	$1 1/4$	$1 1/2$	$1 7/8$	$2 1/8$	$2 5/8$	$3 3/16$	$3 7/8$	$4 7/16$	5	$6 1/4$	$7 3/8$

Above the numeric columns: Nominal or trade size of conduit, inches

Figure A–9   Approximate dimensions of standard locknuts and bushings. (Courtesy of American Iron and Steel Institute.)

Recommended Minimum Spacing of Rigid Steel Conduit and EMT at Junction and Pull Boxes

Distances between centers, inches

Nominal or trade size inches	$1/2$	$3/4$	1	$1 1/4$	$1 1/2$	2	$2 1/2$	3	$3 1/2$	4	5	6
1/2	$1 3/8$	—	—	—	—	—	—	—	—	—	—	—
3/4	$1 1/2$	$1 5/8$	—	—	—	—	—	—	—	—	—	—
1	$1 3/4$	$1 7/8$	2	—	—	—	—	—	—	—	—	—
1 1/4	2	$2 1/8$	$2 1/4$	$2 1/2$	—	—	—	—	—	—	—	—
1 1/2	$2 1/8$	$2 1/4$	$2 3/8$	$2 5/8$	$2 3/4$	—	—	—	—	—	—	—
2	$2 3/8$	$2 1/2$	$2 3/4$	3	$3 1/8$	$3 3/8$	—	—	—	—	—	—
2 1/2	$2 5/8$	$2 3/4$	3	$3 1/4$	$3 3/8$	$3 5/8$	4	—	—	—	—	—
3	3	$3 1/8$	$3 3/8$	$3 5/8$	$3 3/4$	4	$4 3/8$	$4 3/4$	—	—	—	—
3 1/2	$3 3/8$	$3 1/2$	$3 5/8$	$3 7/8$	4	$4 3/8$	$4 5/8$	5	$4 3/4$	—	—	—
4	$3 11/16$	$3 7/8$	4	$4 1/4$	$4 3/8$	$4 3/4$	5	$5 3/8$	$5 5/8$	6	—	—
5	$4 3/8$	$4 1/2$	$4 5/8$	$4 7/8$	5	$5 3/8$	$5 5/8$	6	$6 1/4$	$6 5/8$	$7 1/4$	—
6	5	$5 1/8$	$5 1/4$	$5 1/2$	$5 5/8$	6	$6 1/4$	$6 5/8$	7	$7 1/4$	8	$8 5/8$

Figure A–10   Recommended spacing for conduit and tubing where run in racks or where entering a panelboard or junction box. (Courtesy of American Iron and Steel Institute.)

DC Circuit Characteristics

Ohm's Law:

$$E = IRI = \frac{E}{R} R = \frac{E}{I}$$

E = voltage impressed on circuit (volts)

I = current flowing in circuit (amperes)

R = circuit resistance (ohms)

Resistances in Series:

$R_t = R_1 + R_2 + R_3 + \dots$

R_T = total resistance (ohms)

R_1, R_2 etc. = individual resistances (ohms)

Resistances in Parallel:

$$R_t = \frac{1}{\frac{1}{R_1} + \frac{1}{R_2} + \frac{1}{R_3} + \dots}$$

Formulas for the conversion of electrical and mechanical power:

$HP = \frac{watts}{746}$ (watts × .00134)

$HP = \frac{kilowatts}{.746}$ (kilowatts × 1.34)

Kilowatts = HP × .746

Watts = HP × 746

HP = Horsepower

In direct-current circuits, electrical power is equal to the product of the voltage and current:

$$P = EI = I_2 R = \frac{E_2}{R}$$

P = power (watts)

E = voltage (volts)

I = current (amperes)

R = resistance (ohms)

Solving the basic formula for I, E, and R gives

$$I = \frac{P}{E} = \sqrt{\frac{P}{R}}; \quad E = \frac{P}{I} = \sqrt{RP}; \quad R = \frac{E_2}{P} = \frac{P}{T^2}$$

Energy

Energy is the capacity for doing work. Electrical energy is expressed in kilowatt-hours (kWhr), one kilowatt-hour representing the energy expended by a power source of 1 kW over a period of 1 hour.

Efficiency

Efficiency of a machine, motor or other device is the ratio of the energy output (useful energy delivered by the machine) to the energy input (energy delivered to the machine), usually expressed as a percentage:

Efficiency = $\frac{output}{input}$ × 100%

or Output = imput × $\frac{efficiency}{100\%}$

Torque

Torque may be described as a force tending to cause a body to rotate. It is expressed in pound-feet or pounds of force acting at a certain radius:

Torque (pound-feet) = force tending to produce rotation (pounds) × distance from center of rotation to point at which force is applied (feet).

Relations between torque and horsepower:

Torque = $\frac{33{,}000 \times HP}{6.28 \times rpm}$

HP = $\frac{6.28 \times rpm \; time \; torque}{33{,}000}$

rpm = speed of rotating part (revolutions per minute)

AC Circuit Characteristics

The instantaneous values of an alternating current or voltage vary from zero to maximum value each half cycle. In the practical formula that follows, the "effective value" of current and voltage is used, defined as follows:

Effective value = 0.707 × maximum instantaneous value

Inductances in Series and Parallel:

The resulting circuit inductance of several inductances in series or parallel is determined exactly as the sum of resistances in series or parallel as described under DC circuit characteristics.

Impedance:

Impedance is the total opposition to the flow of alternating current. It is a function of resistance, capacitive reactance and inductive reactance. The following formulae relate these circuit properties:

$$X_L = 2\pi Hz L \quad X_c = \frac{1}{2\pi Hz C} \quad Z = \sqrt{R^2 + (X_L - X_C)^2}$$

X_L = inductive reactance (ohms)

X_c = capacitive reactance (ohms)

Z = impedance (ohms)

Hz - (Hertz) cycles per second

C = capacitance (farads)

L - inductance (henrys)

R = resistance (ohms)

$\pi = 3.14$

In circuits where one or more of the properties L, C, or R is absent, the impedance formula is simplified as follows:

Resistance only:	Inductance only:	Capacitance only:
Z = R	$Z = X_L$	Z = XC
Resistance and Inductance only:	Resistance and Capacitance only:	Inductance and Capacitance only:
$Z = \sqrt{R^2 + X_L^2}$	$Z = \sqrt{R^2 + X_C^2}$	$Z = \sqrt{X_L - X_C^2}$

Ohm's law for AC circuits:

$$E = 1 \times Z \qquad I = \frac{E}{Z} \qquad Z = \frac{E}{I}$$

Capacitances in Parallel:

$C_t = C_1 + C_2 \; C_3 + \dots$

C_t = total capacitance (farads)

$C_1 C_2 C_3 \dots$ = individual cpacitances (farads)

Figure A–11(A) Common electrical formulas for making AC and DC calculations. (Courtesy of American Iron and Steel Institute.)

Appendix

Capacitances in series:

$$C_t = \frac{1}{\frac{1}{C_1} + \frac{1}{C_2} + \frac{1}{C_3} + \ldots}$$

Phase Angle

An alternating current through an inductance lags the voltage across the inductance by an angle computed as follows:

Tangent of angle of lag $= \frac{X_L}{R}$

An alternating current through a capacitance leads the voltage across the capacitance by an angle computed as follows:

Tangent of angle of lead $= \frac{X_C}{R}$

The resultant angle by which a current leads or lags the voltage in an entire circuit is called the phase angle and is computed as follows:

Cosine of phase angle $= \frac{R \text{ of circuit}}{Z \text{ of circuit}}$

Power Factor

Power factor of a circuit or system is the ratio of actual power (watts) to apparent power (volt-amperes), and is equal to the cosine of the phase angle of the circuit:

$$PF = \frac{\text{actual power}}{\text{apparent power}} = \frac{\text{watts}}{\text{volts} \times \text{amperes}} = \frac{kW}{kVA} = \frac{R}{Z}$$

KW = kilowatts
kVA = kilowatt-amperes = volt-amperes + 1,000
PF = power factor (expressed as decimal or percent)

Single-Phase Circuits

$$kVA = \frac{EI}{1,000} = \frac{kW}{PF} \quad kW = kVA \times PF$$

$$I = \frac{P}{E \times PF} \quad E = \frac{P}{I \times PF} \quad PF = \frac{P}{E \times I}$$

$P = E \times I \times PF$
$P = \text{power (watts)}$

Two-Phase Circuits

$$I = \frac{P}{2 \times E \times PF} \quad E = \frac{P}{2 \times I \times PF} \quad PF = \frac{P}{E \times I}$$

$$kVA = \frac{2 \times E \times I}{1000} = \frac{kW}{PF} \quad kW = kVA \times PF$$

$P = 2 \times E \times 1 \times PF$
$E = \text{phase voltage (volts}$

Three-Phase Circuits, Balanced Star or Wye

$I_N = O \quad I = I_P \quad E = \sqrt{3}E_P = 1.73E_P$

$$E_P = \frac{E}{\sqrt{3}} = \frac{E}{1.73} = 0.577E$$

I_N = current in neutral (amperes)
I = line current per phase (amperes)
I_P = current in each phase winding (amperes)
E = voltage, phase to phase (volts)
E_P = voltage, phase to neutral (volts)

Three-Phase Circuits, Balanced Delta

$I = 1.732 \times I_P \quad I_P = \frac{1}{\sqrt{3}} = 0.577 \times I$

$E = E_P$

Power:

Balanced 3-Wire, 3-Phase Circuit, Delta or Wye

For unit power factor (PF = 1.0):

$P = 1.732 \times E \times I$

$I = \frac{P}{\sqrt{3}} \quad E = 0.577\frac{P}{E} \qquad E = \frac{P}{\sqrt{3}} \times I = 0.577\frac{P}{I}$

P = total power (watts)

For any load:

$P = 1.732 \times E \times I \times PF \quad VA = 1.732 \times E \times I$

$E = \frac{P}{PF \times 1.73 \times I} = 0.577 \times \frac{P}{PF \times I}$

$I = \frac{P}{PF \times 1.73 \times E} = 0.577 \times \frac{P}{I} \times E$

$PF = \frac{P}{1.73 \times I \times E} = \frac{0.577 \times P}{I \times E}$

VA = apparent power (volt-amperes)
P = actual power (watts)
E = line voltage (volts)
I = line current (amperes)

Power Loss:

Any AC or DC Circuit

$P = I_2RI = \sqrt{\frac{P}{R}}R = \frac{P}{I^2}$

P = power heat loss in circuit (watts)
I = effective current in conductor (amperes)
R = conductor resistance (ohms)

Load Calculations

Branch Circuits—Lighting & Appliance 2-Wire:

$I = \frac{\text{total connected load (watts)}}{\text{line voltage (volts)}}$

I = current load on conductor (amperes)

3-Wire:

Apply same formula as for 2–wire branch circuit, considering each line to neutral separately. Use line-to-neutral voltage; result gives current in line conductors.

Figure A–11(B) Common electrical formulas for making AC and DC calculations. (Courtesy of American Iron and Steel Institute.)

Electrical Wiring Symbols
Selected from American National Standard Graphic for
Electrical Wiring and Layout Diagrams Used in Architecture and Building Construction
ANSI Y32.9-1972

1. Lighting Outlets

Ceiling *Wall*

1.1 Surface or pendant incandescent, mercury-vapor, or similar lamp fixture

1.2 Recessed incandescent, mercury-vapor or similar lamp fixture

1.3 Surface of pendant individual fluorescent fixture

1.4 Recessed individual fluorescent fixture

1.5 Surface or pendant continuous-row fluorescent fixture

1.6 Recessed continuous-row fluorescent fixture

1.8 Surface or pendant exit light

1.9 Recessed exit light

1.10 Blanked outlet

1.11 Junction box

1.12 Outlet controlled by low-voltage switching when relay is installed in outlet box

2. Receptacle Outlets

Grounded *Ungrounded*

2.1 Single receptacle outlet

2.2 Duplex receptacle outlet

2.3 Triplex receptacle outlet

2.4 Quadrex receptacle outlet

2.5 Duplex receptacle outlet – split wired

2.6 Triplex receptacle outlet – split wired

2.7 Single special-purpose receptacle outlet – split wired

NOTE 2.7A: Use numeral or letter as a subscript alongside the symbol, keyed to explanation in the drawing list of symbols, to indicate type of receptacle or usage.

2.8 Duplex special-purpose receptacle outlet
See note 2.7A

Figure A–12(A)   Common electrical symbols. (Courtesy of American Iron and Steel Institute.)

2.9 Range outlet (typical)
See note 2.7A

2.10 Special-purpose connection or provision
for connection
Use subscript letters to indicate function
(SW – dishwasher; CD – clothes dryer, etc).

UNG
DW

2.12 Clock hanger receptacle

UNG

2.13 Fan hanger receptacle

UNG

2.14 Floor single receptacle outlet

UNG

2.15 Floor duplex receptacle outlet

UNG

2.16 Floor special-purpose outlet
See note 2.7A

UNG

2.17 Floor telephone outlet – public

2.18 Floor telephone outlet – private

2.19 Underfloor duct and junction box for
triple, double, or single duct system (as
indicated by the number of parallel lines)

2.20 Cellular floor header duct

3. Switch Outlets

3.1 Single-pole switch

S

3.2 Double-pole switch

S_2

3.3 Three-way switch

S_3

3.4 Four-way switch

S_4

3.5 Key-operated switch

S_K

3.6 Switch and pilot lamp

S_P

3.7 Switch for low-voltage switching system

S_L

3.8 Master switch for low-voltage switching system

S_{LM}

3.9 Switch and single receptacle

S

3.10 Switch and double receptacle

S

3.11 Door switch

S_D

3.12 Time switch

S_T

3.13 Circuit breaker switch

S_{CB}

Figure A–12(B) Common electrical symbols. (Courtesy of American Iron and Steel Institute.)

3.14 Momentary contact switch or pushbutton
for other than signalling system

SMC

3.15 Ceiling pull switch

Ⓢ

5.13 Radio outlet

R

5.14 Television outlet

TV

6. Panelboards, switchboards, and related equipment

6.1 Flush-mounted panel board and cabinet
NOTE 6.1A: Identify by notation or schedule

6.2 Surface-mounted panel board and cabinet
See note 6.1A

6.3 Switchboard, power control center, unit
substations (should be drawn to scale)
See note 6.1A

6.4 Flush-mounted terminal cabinet
See note 6.1A

NOTE 6.4A: In small-scale drawings theTC may be
indicated alongside the symbol

TC

6.5 Surface-mounted terminal cabinet
See note 6.1A and 6.4A

TC

6.6 Pull box
Identify in relation to wiring system section and size

6.7 Motor or other power controller
See note 6.1A

MC

6.8 Externally operated disconnection switch
See note 6.1A

6.9 Combination controller and disconnection means
See note 6.1A

7. Bus Ducts and Wireways

7.1 Trolley duct
See note 6.1A

| T | | T | | T |

7.2 Busway (service, feeder, or plug-in)
See note 6.1A

| B | | B | | B |

7.3 Cable through, ladder, or channel
See note 6.1A

| BP | | BP | | BP |

7.4 Wireway
See note 6.1A

| W | | W | | W |

9. Circuiting

Wiring method indentification by notation on
drawing or in specifications.

9.1 Wiring concealed in ceiling or wall

NOTE 9.1A: Use heavy weight line to identify service
and feed runs.

9.2 Wiring concealed in floor
See note 9.1A

Figure A–12(C)   Common electrical symbols. (Courtesy of American Iron and Steel Institute.)

9.2 Wiring exposed
  See note 9.1A

_ _ _ _ _ _ _ _ _ _

9.4 Branch circuit home run to panel board
  Number of arrows indicates number of circuits. (A numeral at each arrow may be used
to identify circuit number.)

2   1

NOTE: Any circuit without further identification indicates a 2-wire circuit. For a greater number of wires,
indicate with cross lines (see 9.4.1, Applications)

9.4.1 Applications:

—///— 3 wires;

—////— 4 wires, etc

Unless indicated otherwise, the wire size of the circuit
is the minimum size required by the specification.
Indicate size in inches and identify different functions
of wiring system, such as signaling, by notation or
other means.

9.6 Wiring turned up

—————————○

9.7 Wiring turned down

—————————●

Figure A–12(D) Common electrical symbols. (Courtesy of American Iron and Steel Institute.)

These environmental resistance ratings are based upon tests where the specimens were placed in complete submergence in the reagent listed. In many applications Carflex* conduit can be used in process areas where these chemicals are manufactured or used because worker safety requirements dictate that any air presence or splashing be at a very low level. Most liquidtight conduit is located in areas suitable for worker access.

If there are any questions for specific suitability in a given environment, prototype samples should be tested under actual conditions.

RATING CODE

A-Excellent service.
No harmful effect to reduce service life. Suitable for continuous service.

B-Good service life.
Moderate to minor effect. Good for intermittent service. Generally suitable for continuous service.

C-Fair or limited service.
Depends on operating conditions. Generally suitable for intermittent continuous service.

D-Unsatisfactory service.
Not recommended.

* These ratings apply only to Carflex® conduit & tubing.

Chemical	Concentration	Resistance
Acetate Solvents		D
Acetic Acid	10%	B
Acetic Acid (Glacial)		C
Acetone		D
Acrylonitrile		A
Alcohols (Aliphatic)		C
Aluminum Chloride		A
Aluminum Sulfate (Alums)		A
Ammonia (Anhydrous Liquids)		D
Ammonia (Aqueous)		A
Ammoniated Latex		A
Ammonium Chloride		A
Ammonium Hydroxide		A
Amyl Acetate		D
Aniline Oils		D
Aromatic Hydrocarbons		D
Asphalt		D
ASTM Fuel A		A
ASTM Fuel B		D
ASTM #1 Oil		B
ASTM #3 Oil		C
Barium Chloride		A
Barium Sulfide		A
Barium Hydroxide		A
Benzene (Benzol)		D
Benzine (Petroleum Ether)		C
Black Liquor		A
Bordeaux Mixture		A
Boric Acid		A
Butyl Acetate		D
Butyl Alcohol		B
Calcium Hydroxide		A
Calcium Hypochlorite		A
Carbolic Acid (Phenol)		B
Carbon Dioxide		A
Carbon Disulfide		D
Carbon Tetrachloride		D
Carbonic Acid		A
Casein		A
Caustic Soda		A
Chlorine Gas (wet)		D
Chlorine Gas (dry)		D
Chlorine (water solution)		C
Chlorobenzene		D
Chlorinated Hydrocarbons		D
Chromic Acid	10%	B
Citric Acid		A
Coal Tar		D
Coconut Oil		C
Corn Oil		A
Cottonseed Oil		C
Creosote		D
Cresol		C
Cresylic Acid		D
Cyclohexane		B
DDT Weed Killer		A
Dibutyl Phthalate		D
Diesel Oils		C
Diethylene Glycol		B
Diethyl Ether		A
Di-isodecyl Phthalate		D
Dioctyl Phthalate		D
Dow General Weed Killer (Phenol)		D
Dow General Weed Killer (H_2O)		B
Ethyl Alcohol		C
Ethylene Dichloride		D
Ethylene Glycol		B
Ferric Chloride		A
Ferric Sulfate		A
Ferrous Chloride		A
Ferrous Sulfate		A
Formaldehyde		D
Fuel Oil		B
Furfural		C
Gallic Acid		A
Gasoline (Hi Test)		C
Glycerine		A
Grease		A
Green Sulfate Liquor		A
Heptachlor in Petroleum Solvents		A
Heptane		C
Hexane		C
Hydrobromic Acid		A
Hydrochloric Acid	10%	A
Hydrochloric Acid	40%	A
Hydrofluoric Acid	70%	D
Hydrofluoroboric Acid		A
Hydrofluorosilicic Acid		A
Hydrogen Peroxide	10%	A
Iso-octane		C
Isopropyl Acetate		D
Isopropyl Alcohol		B
Jet Fuels (JP-3, 4, and 5)		C
Kerosene		C
Ketones		D
Linseed Oil		A
Lubricating Oils		A
Magnesium Chloride		A
Magnesium Hydroxide		A
Magnesium Sulfate		A
Malathion 50 in Aromatics		D
Malic Acid		A
Methyl Acetate		D
Methyl Alcohol		C
Methyl Bromide		D
Methyl Ethyl Ketone		D
Methylene Chloride		D
Mineral Oil		A
Monochlorobenzene		D
Muriatic Acid (See Hydrochloric Acid)		—
Naphtha		C
Naphthalene		D
Nitric Acid	10%	A
Nitric Acid	35%	A
Nitric Acid	70%	D
Oleic Acid		A
Oleum		D
Oxalic Acid		A
Pentachlorophenol in Oil		B
Pentane		C
Perchloroethylene		D
Petroleum Ether		C
Phenol		B
Phosphoric Acid	85%	A
Pitch		B
Potassium Hydroxide		A
Propyl Alcohol		B
Ritchfield "A" Weed Killer		C
Sea Water		A
Sodium Hydroxide	10%	A
Sodium Hydroxide	50%	A
Soybean Oil		C
Sodium Cyanide		A
Stoddard Solvent		D
Styrene		D
Sulfur Dioxide (liquid)		D
Sulfuric Acid	50%	A
Sulfuric Acid	98%	D
Sulfurous Acid		B
Tall Oil		D
Tannic Acid		A
Toluene		D
Trichlorethylene		D
Triethanol Amine		C
Tricresyl Phosphate (Skydrol)		D
Turpentine		C
Vinegar		A
Vinyl Chloride		D
Water		A
White Liquor		A
Xylene		D
Zinc Chloride		A
Zinc Sulfate		A

Figure A–13 Chemical resistance chart and recommended rating code for PVC electrical products with specific liquids and dusts. (Courtesy of Carlon, a Lampson Sessions Company.)

Urethane Interior Coating Chemical Resistance Chart

Solutions	Conc.	Temp.	Splashing	Liquid	Fumes
Acetic Acid	10%	75	yes	yes	yes
Acid Copper Plating Solution		75	yes	yes	yes
Alkaline Cleaners		75	yes	yes	yes
Aluminum Chloride	Sat'd	75	yes	yes	yes
Aluminum Sulfate	Sat'd	75	yes	yes	yes
Alums	Sat'd	75	yes	yes	yes
Ammonium Chloride	Sat'd	75	yes	yes	yes
Ammonium Hydroxide	28%	75	yes	yes	yes
Ammonium Hydroxide	10%	75	yes	yes	yes
Ammonium Sulfate	Sat'd	75	yes	yes	yes
Ammonium Thiocyanate	Sat'd	75	yes	yes	yes
Amyl Alcohol	Any	75	?	no	?
Arsenic Acids	Any	75	yes	yes	yes
Barium Sulfide	Sat'd	75	yes	yes	yes
Black Liquor	Sat'd	75	yes	yes	yes
Benzoic Acid	Sat'd	75	yes	yes	yes
Brass Plating Solution	Any	75	yes	yes	yes
Bromine Water	Sat'd	75	yes	no	yes
Butyl Alcohol	Any	75	?	no	?
Cadmium Plating Solution	Any	75	yes	yes	yes
Calcium Bisulfite	Any	75	yes	yes	yes
Calcium Chloride	Sat'd	75	yes	yes	yes
Calcium Hypochlorite	Sat'd	75	yes	yes	yes
Carbonic Acid	Sat'd	75	yes	yes	yes
Casein	Sat'd	75	yes	yes	yes
Castor Oil	Any	75	yes	yes	yes
Caustic Soda	35%	75	yes	yes	yes
Caustic Soda	10%	75	yes	yes	yes
Caustic Potash	35%	75	yes	yes	yes
Caustic Potash	10%	75	yes	yes	yes
Chlorine Water	Sat'd	75	yes	yes	yes
Chromium Plating Solution	Any	75	yes	yes	yes
Citric Acid	Sat'd	75	yes	yes	yes
Copper Chloride (Cupric)	Sat'd	75	yes	yes	yes
Copper CyanidePlating Sol (High Speed)	Any	75	yes	yes	yes
(High Speed)	Any	75	yes	yes	yes
(with Alkali Cyanides)	Sat'd	75	yes	yes	yes
Copper Sulfate	Sat'd	75	yes	yes	yes
Cocoanut Oil	Sat'd	75	yes	yes	yes
Cottonseed Oil	Sat'd	75	yes	yes	yes
Disodium Phosphate	Sat'd	75	yes	yes	yes
Ethyl Alcohol	Any	75	?	no	?
Ethylene Glycol	Any	75	?	no	?
Ferric Chloride	45%	75	yes	yes	yes
Ferrous Sulfate	Sat'd	75	yes	yes	yes
Fluoboric Acid	Any	75	yes	yes	yes
Formaldehyde	37%	75	yes	yes	yes
Formic Acid	85%	75	yes	no	yes
Gallic Acid	Sat'd	75	yes	yes	yes
Glucose	Any	75	yes	yes	yes
Glue	Any	75	yes	yes	yes
Glycerine	Any	75	yes	yes	yes
Gold Plating Solution	Any	75	yes	yes	yes
Hydrochloric Acid	10%	75	yes	yes	yes
Hydrochloric Acid	21.5%	75	yes	yes	yes
Hydrochloric Acid	37.5%	75	yes	no	yes
Hydrofluoric Acid	4%	75	yes	yes	yes
Hydrofluoric Acid	10%	75	yes	yes	yes
Hydrofluoric Acid	48%	75	yes	no	yes
Hydrogen Peroxide	30%	75	yes	yes	yes
Hydrogen Sulfide	Sat'd	75	yes	yes	yes
Hydroquinone	Any	75	yes	yes	yes
Indium Plating Solution	Any	75	yes	yes	yes
Lactic Acid	50%	75	yes	yes	yes
Lactic Acid	Any	75	yes	yes	yes
Lead Plating Solution	Any	75	yes	yes	yes
Malic acid	Any	75	yes	yes	yes
Methyl Alcohol	Any	75	?	no	?
Mineral Oils	Any	75	yes	yes	yes
Nickel Acetate	Sat'd	75	yes	yes	yes
Nickel Plating Solution		75	yes	yes	yes
Nickel Salts	Sat'd	75	yes	yes	yes
Nitric Acid	35%	75	yes	yes	yes
Nitric Acid	40%	75	yes	yes	yes
Nitric Acid	60%	75	yes	no	yes
Nitric Acid/ Hydrofluoric Acid	15 % 4%	75	yes	yes	yes
Nitric Acid/ Sodium Dichromate Water	16 % 13% 71%	75	yes	yes	yes
Oleic Acid	Any	75	yes	yes	yes
Oxalic Acid	Sat'd	75	yes	yes	yes
	Any	75	yes	yes	yes
Phenol	Sat'd	75	yes	no	yes
Phosphoric Acid	75%	75	yes	yes	yes
Phosphoric Acid	85%	75	yes	yes	yes
Potassium Acid Sulfate	Sat'd	75	yes	yes	yes
Potassium Antimonate	Sat'd	75	yes	yes	yes
Potassium Bisulfite	Sat'd	75	yes	yes	yes
Potassium Chloride	Sat'd	75	yes	yes	yes
Potassium Cuprocyanide	Sat'd	75	yes	yes	yes
Potassium Cyanide	Sat'd	75	yes	yes	yes
Potassium Diachromate	Sat'd	75	yes	yes	yes
Potassium Hypochlorite	Sat'd	75	yes	?	yes
Potassium Sulfide	Sat'd	75	yes	yes	yes
Potassium Thiosulfate	Sat'd	75	yes	yes	yes
Propyl Alcohol	Sat'd	75	?	no	?
Rhodium Plating Solution	Sat'd	75	yes	yes	yes
Silver Plating Solution	Sat'd	75	yes	yes	yes
Soaps	Any	75	yes	yes	yes
Sodium Acid Sulfate	Sat'd	75	yes	yes	yes
Sodium Antimonate	Sat'd	75	yes	yes	yes
Sodium Bicarbonate	Sat'd	75	yes	yes	yes
Sodium Bisulfite	Sat'd	75	yes	yes	yes
Sodium Chloride	Sat'd	75	yes	yes	yes
Sodium Cyanide	Sat'd	75	yes	yes	yes
Sodium Dichromate	Sat'd	75	yes	yes	yes
Sodium Hydroxide	10%	75	yes	yes	yes
Sodium Hydroxide	35%	75	yes	yes	yes
Sodium Hydroxide	73%	75	no	no	yes
Sodium Hypochlorite	Sat'd	75	yes	?	yes
Sodium Hypochlorite	15%	75	yes	?	yes
Sodium Sulfide	Sat'd	75	yes	yes	yes
Sodium Thiosulfate	Sat'd	75	yes	yes	yes
Sulfuric Acid	15%	75	yes	yes	yes
Sulfuric Acid	50%	75	yes	yes	yes
Sulfuric Acid	70%	75	yes	yes	yes
Sulfuric Acid	98%	75	no	no	yes
Sulfurous Acid	2%	75	yes	yes	yes
Sulfurous Acid	6%	75	yes	no	yes
Tannic Acid	Sat'd	75	yes	yes	yes
Tartaric Acid	Sat'd	75	yes	yes	yes
Tin Chloride Aqueous	Sat'd	75	yes	yes	yes
Tin Plating Solution	Sat'd	75	yes	yes	yes
Triethaneolamine	Sat'd	75	yes	yes	yes
Trisodium Phosphate	Sat'd	75	yes	yes	yes
Water	Sat'd	75	yes	yes	yes
White Liquor		75	yes	yes	yes
Zinc Plating Solution		75	yes	yes	yes
Zinc Sulfate	Sat'd	75	yes	yes	yes

Figure A–14 Chemical resistance chart for Urethane coatings. (Courtesy of Robroy Industries.)

184

PVC Coating Chemical Resistance Chart

Solutions	Conc.	Temp.	Splashing	Liquid	Fumes
Acetic Acid	10%	75	yes	yes	yes
Acid Copper Plating Solution		75	yes	yes	yes
Alkaline Cleaners		75	yes	yes	yes
Aluminum Chloride	Sat'd	75	yes	yes	yes
Aluminum Sulfate	Sat'd	75	yes	yes	yes
Alums	Sat'd	75	yes	yes	yes
Ammonium Chloride	Sat'd	75	yes	yes	yes
Ammonium Hydroxide	28%	75	yes	yes	yes
Ammonium Hydroxide	10%	75	yes	yes	yes
Ammonium Sulfate	Sat'd	75	yes	yes	yes
Ammonium Thiocyanate	Sat'd	75	yes	yes	yes
Amyl Alcohol	Any	75	?	no	?
Arsenic Acids	Any	75	yes	yes	yes
Barium Sulfide	Sat'd	75	yes	yes	yes
Black Liquor	Sat'd	75	yes	yes	yes
Benzoic Acid	Sat'd	75	yes	yes	yes
Brass Plating Solution	Any	75	yes	yes	yes
Bromine Water	Sat'd	75	yes	no	yes
Butyl Alcohol	Any	75	?	no	?
Cadmium Plating Solution	Any	75	yes	yes	yes
Calcium Bisulfite	Any	75	yes	yes	yes
Calcium Chloride	Sat'd	75	yes	yes	yes
Calcium Hypochlorite	Sat'd	75	yes	yes	yes
Carbonic Acid	Sat'd	75	yes	yes	yes
Casein	Sat'd	75	yes	yes	yes
Castor Oil	Any	75	yes	yes	yes
Caustic Soda	35%	75	yes	yes	yes
Caustic Soda	10%	75	yes	yes	yes
Caustic Potash	35%	75	yes	yes	yes
Caustic Potash	10%	75	yes	yes	yes
Chlorine Water	Sat'd	75	yes	yes	yes
Chromium Plating Solution	Any	75	yes	yes	yes
Citric Acid	Sat'd	75	yes	yes	yes
Copper Chloride (Cupric)	Sat'd	75	yes	yes	yes
Copper CyanidePlating Sol (High Speed)	Any	75	yes	yes	yes
	Any	75	yes	yes	yes
(with Alkali Cyanides)	Sat'd	75	yes	yes	yes
Copper Sulfate	Sat'd	75	yes	yes	yes
Cocoanut Oil	Sat'd	75	yes	yes	yes
Cottonseed Oil	Sat'd	75	yes	yes	yes
Disodium Phosphate	Sat'd	75	yes	yes	yes
Ethyl Alcohol	Any	75	?	no	?
Ethylene Glycol	Any	75	?	no	?
Ferric Chloride	45%	75	yes	yes	yes
Ferrous Sulfate	Sat'd	75	yes	yes	yes
Fluoboric Acid	Any	75	yes	yes	yes
Formaldehyde	37%	75	yes	yes	yes
Formic Acid	85%	75	yes	no	yes
Gallic Acid	Sat'd	75	yes	yes	yes
Glucose	Any	75	yes	yes	yes
Glue	Any	75	yes	yes	yes
Glycerine	Any	75	yes	yes	yes
Gold Plating Solution	Any	75	yes	yes	yes
Hydrochloric Acid	10%	75	yes	yes	yes
Hydrochloric Acid	21.5%	75	yes	yes	yes
Hydrochloric Acid	37.5%	75	yes	no	yes
Hydrofluoric Acid	4%	75	yes	yes	yes
Hydrofluoric Acid	10%	75	yes	yes	yes
Hydrofluoric Acid	48%	75	yes	no	yes
Hydrogen Peroxide	30%	75	yes	yes	yes
Hydrogen Sulfide	Sat'd	75	yes	yes	yes
Hydroquinon	Any	75	yes	yes	yes
Indium Plating Solution	Any	75	yes	yes	yes
Lactic Acid	50%	75	yes	yes	yes
Lactic Acid	Any	75	yes	yes	yes
Lead Plating Solution	Any	75	yes	yes	yes
Malic acid	Any	75	yes	yes	yes
Methyl Alcohol	Any	75	?	no	?
Mineral Oils	Any	75	yes	yes	yes
Nickel Acetate	Sat'd	75	yes	yes	yes
Nickel Plating Solution		75	yes	yes	yes
Nickel Salts	Sat'd	75	yes	yes	yes
Nitric Acid	35%	75	yes	yes	yes
Nitric Acid	40%	75	yes	yes	yes
Nitric Acid	60%	75	yes	no	yes
Nitric Acid/	15%				
Hydrofluoric Acid	4%	75	yes	yes	yes
Nitric Acid/	16%				
Sodium Dichromate	13%	75	yes	yes	yes
Water	71%				
Oleic Acid	Any	75	yes	yes	yes
Oxalic Acid	Sat'd	75	yes	yes	yes
	Any	75	yes	yes	yes
Phenol	Sat'd	75	yes	no	yes
Phosphoric Acid	75%	75	yes	yes	yes
Phosphoric Acid	85%	75	yes	yes	yes
Potassium Acid Sulfate	Sat'd	75	yes	yes	yes
Potassium Antimonate	Sat'd	75	yes	yes	yes
Potassium Bisulfite	Sat'd	75	yes	yes	yes
Potassium Chloride	Sat'd	75	yes	yes	yes
Potassium Cuprocyanide	Sat'd	75	yes	yes	yes
Potassium Cyanide	Sat'd	75	yes	yes	yes
Potassium Diachromate	Sat'd	75	yes	yes	yes
Potassium Hypochlorite	Sat'd	75	yes	?	yes
Potassium Sulfide	Sat'd	75	yes	yes	yes
Potassium Thiosulfate	Sat'd	75	yes	yes	yes
Propyl Alcohol	Sat'd	75	?	no	?
Rhodium Plating Solution	Sat'd	75	yes	yes	yes
Silver Plating Solution	Sat'd	75	yes	yes	yes
Soaps	Any	75	yes	yes	yes
Sodium Acid Sulfate	Sat'd	75	yes	yes	yes
Sodium Antimonate	Sat'd	75	yes	yes	yes
Sodium Bicarbonate	Sat'd	75	yes	yes	yes
Sodium Bisulfite	Sat'd	75	yes	yes	yes
Sodium Chloride	Sat'd	75	yes	yes	yes
Sodium Cyanide	Sat'd	75	yes	yes	yes
Sodium Dichromate	Sat'd	75	yes	yes	yes
Sodium Hydroxide	10%	75	yes	yes	yes
Sodium Hydroxide	35%	75	yes	yes	yes
Sodium Hydroxide	73%	75	no	no	yes
Sodium Hypochlorite	Sat'd	75	yes	?	yes
Sodium Hypochlorite	15%	75	yes	?	yes
Sodium Sulfide	Sat'd	75	yes	yes	yes
Sodium Thiosulfate	Sat'd	75	yes	yes	yes
Sulfuric Acid	15%	75	yes	yes	yes
Sulfuric Acid	50%	75	yes	yes	yes
Sulfuric Acid	70%	75	yes	yes	yes
Sulfuric Acid	98	75	no	no	yes
Sulfurous Acid	2%	75	yes	yes	yes
Sulfurous Acid	6%	75	yes	no	yes
Tannic Acid	Sat'd	75	yes	yes	yes
Tartaric Acid	Sat'd	75	yes	yes	yes
Tin Chloride Aqueous	Sat'd	75	yes	yes	yes
Tin Plating Solution	Sat'd	75	yes	yes	yes
Triethaneolamine	Sat'd	75	yes	yes	yes
Trisodium Phosphate	Sat'd	75	yes	yes	yes
Water	Sat'd	75	yes	yes	yes
White Liquor		75	yes	yes	yes
Zinc Plating Solution		75	yes	yes	yes
Zinc Sulfate	Sat'd	75	yes	yes	yes

Figure A–15   Chemical resistance chart for PVC coatings. (Courtesy of Robroy Industries.)

INDEX